Builder's Guide to Mixed Climates

Details for Design and Construction

Joseph Lstiburek, P. Eng

The Taunton Press

Printed in the United States of America
10 9 8 7 6 5 4 3 2 1

For Pros / By Pros™ is a trademark of The Taunton Press, Inc., registered in the U.S. Patent and Trademark Office.

The Taunton Press, Inc., 63 South Main Street, PO Box 5506, Newtown, CT 06470-5506
e-mail: tp@taunton.com

Distributed by Publishers Group West

Previously published as *Builder's Guide to Mixed-Humid Climates* by Building Science Corporation, 70 Main Street, Westford, MA 01886, (978) 589-5100
www.buildingscience.com

Taunton
BOOKS & VIDEOS

for fellow enthusiasts

Library of Congress Cataloging-in-Publication Data
Lstiburek, Joseph W.
For Pros / By Pros™: The builder's guide to mixed climates : details for design and construction / Joseph Lstiburek.
p.   cm.
ISBN 1-56158-388-X
1. Building. 2. Architecture and climate. 3. House construction. I. Title
TH146.L78 2000
693.8—dc21   99-059397

Book Design and Illustrations by:	Sanders/Tikkanen Design & Illustration Bruce Sanders, Ellen Tikkanen, Matthew Harless Westford, MA (978) 692-9800
Cover Photographs:	Front cover: Steve Culpepper/*Fine Homebuilding* magazine, © The Taunton Press, Inc. (bottom left and center); Andrew Wormer/*Fine Homebuilding* magazine, © The Taunton Press, Inc. (bottom right); Back cover: Bruce Greenlaw/*Fine Homebuilding* magazine, © The Taunton Press, Inc.
Additional Illustrations and Production by:	Stephanie Menegus, Steve Baczek Building Science Corporation
Appendix IV Written and Illustrated by:	John Carmody, Building Research Center, University of Minnesota

Comments and constructive criticism of this publication are welcomed by the author and all such comments will be considered in future revisions. Please contact the author directly at joe@buildingscience.com

The information contained in this publication represents or is based upon the viewpoint and understanding of Joseph Lstiburek, Building Science Corporation, and does not necessarily represent the viewpoint and understanding of any other person or entity.

Acknowledgments

The building science information presented in this guide and the companion guides for other climates has evolved over the past 100 years. Many individuals and numerous institutions, organizations and agencies have contributed significantly, often anonymously, to the wealth of information and experience contained here. Fundamental research from the research establishments of four nations — Canada, Norway, Sweden and the United States — provides the foundation for the construction details and approaches presented. More significantly, the lessons learned from the construction of the Arkansas House, Saskatchewan Conservation House, Leger House, Canada's R-2000 program and the U.S. Building America Program were applied, altered, improved, discarded, rediscovered and massaged by EEBA members throughout North America in thousands of field tests and experiments, sometimes referred to as home-building. The experience from these lessons can be found in the following pages.

This guide was significantly enhanced by the thoughtful review of the following individuals whose input is greatly appreciated:

Steve Baczek, AIA, Building Science Corporation
Steve Braun, NAIMA
Dennis Creech, Southface Energy Institute
Rick Graham, Air-Right Energy Design
Michael Guy, Virginia Power
Stephanie Menegus, Building Science Corporation
Neil Moyer, Florida Solar Energy Center
Brad Oberg, IBACOS
Betsy Pettit, AIA, Building Science Corporation
Dick Tracey, D & R International

Joseph Lstiburek,
Westford, MA
April 1997

Revised January 2000

· ·

When we build, let us think that we build forever. Let it not be for present delight nor for present use alone. Let it be such work as our descendants will thank us for; and let us think, as we lay stone on stone, that a time is to come when those stones will be held sacred because our hands have touched them, and that people will say, as they look upon the labor and wrought substance of them, "See! This our parents did for us."

John Ruskin

About this Guide

This guide contains information that is applicable to mixed-humid climates. A mixed-humid climate is defined as a region that receives more than 20 inches of annual precipitation, has approximately 4,500 heating degree days or less and where the monthly average outdoor temperature drops below 45°F during the winter months.

Figure A - Climate Zones illustrates the major climate zones in North America used to distinguish the range of applicability of this guide and the companion guides for cold, hot-humid, mixed-dry and hot-dry climates. Each climate zone specified is broad and general for simplicity. The climate zones are generally based on Herbertson's Thermal Regions, a modified Koppen classification (see Goode's World Atlas, 19th Edition, Rand McNally & Company, New York, NY, 1990), the ASHRAE definition of hot, humid climates (see ASHRAE Fundamentals, ASHRAE, Atlanta, GA, 1997) and average annual precipitation from the U.S. Department of Agriculture. For a specific location, designers and builders should consider weather records, local experience, and the micro climate around a building. Incident solar radiation, nearby water and wetlands, vegetation, and undergrowth can all affect the micro climate.

Although this guide provides general recommendations with applicability based on Figure A, local experience and local building codes should also be considered. Where a conflict between local code and regulatory requirements and the recommendations in this guide occur, authorities having jurisdiction should be consulted or the local code and regulatory requirements should govern.

Illustrations depicting wood framing are shown with exterior walls framed using 2x6 framing techniques. The colored lines or colored shading on illustrations represent materials that form the air flow retarder system.

Precise specification of materials and products is not typically provided on the illustrations or in the text to provide maximum flexibility. It is the responsibility of the designer, builder, supplier and manufacturer to determine specific material compatibility and appropriateness of use. For example, there are wide range of performance and cost issues dealing with sealants, adhesives, tapes and gaskets. Hot weather or cold weather construction and oily, damp or dusty surfaces affect performance along with substrate compatible issues. Tapes must be matched to substrates. Similarly, sealants and adhesives must be matched to materials and joint geometry.

Generally, several different tapes, sealants, adhesives or gaskets can be found to provide satisfactory performance when installed in the locations illustrated in this guide. Premium tapes, sealants, adhesives or gaskets typically (but not always) outperform budget tapes, sealants, adhesives or gaskets. It is always advisable to obtain samples and test compatibility and performance on actual material substrates prior to construction and over an extended period of time.

Figure A
Climate Zones
- Based on Herbertson's Thermal Regions, a modified Koppen Classification, the ASHRAE definition of hot-humid climates and average annual precipitation from the U.S. Department of Agriculture and Environment Canada

Legend

Severe-Cold

A severe-cold climate is defined as a region with approximately 8,000 heating degree days or greater

Cold

A cold climate is defined as a region with approximately 4,500 heating degree days or greater and less than approximately 8,000 heating degree days

Mixed-Humid

A mixed-humid climate is defined as a region that receives more than 20 inches of annual precipitation, has approximately 4,500 heating degree days or less and where the monthly average outdoor temperature drops below 45°F during the winter months

Hot-Humid

A hot-humid climate is defined as a region that receives more than 20 inches of annual precipitation and where the monthly average outdoor temperature remains above 45°F throughout the year*

Hot-Dry/Mixed-Dry

A hot-dry climate is defined as a region that receives less than 20 inches of annual precipitation and where the monthly average outdoor temperature remains above 45°F throughout the year;

A mixed-dry climate is defined as a region that receives less than 20 inches of annual precipitation, has approximately 4,500 heating degree days or less and where the monthly average outdoor temperature drops below 45°F during the winter months

---

* This definition characterizes a region that is almost identical to the ASHRAE definition of hot-humid climates where one or both of the following occur:

- a 67°F or higher wet bulb temperature for 3,000 or more hours during the warmest six consecutive months of the year; or

- a 73°F or higher wet bulb temperature for 1,500 or more hours during the warmest six consecutive months of the year

Contents

Contents

Contents

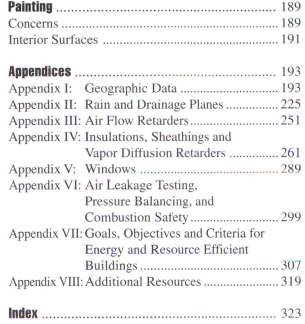

Introduction

People ask builders all the time, "Why can't you build homes the way you used to?" Builders can't help but notice that callbacks and warranty claim costs are rising more rapidly than they did in the past. Were older homes built better than we build today? Have we forgotten all we learned about quality and durability?

Many of the homes of yesterday do perform better than homes built today, but not for the reasons most people think.

Many builders today are obsessed with quality control. The quality revolution that has swept through North America has forced builders to focus on producing quality homes. When faced with the question of quality, a builder typically looks at two things:

- workmanship
- materials

It is assumed that if good workmanship can be achieved and good materials are used, a high quality home will result. Let's examine that assumption.

Countless homes across North America are being built with good workmanship and good materials but are not performing. We have better paints than we ever had before, but we have more paint problems today. Are today's painting contractors less skilled than yesterday's? We have better insulation than we ever had before, but we have more insulation problems today. Are today's insulation contractors that much worse than yesterday's? We have better windows than we ever had before, but we have more window problems today. Are today's window installers that much worse than yesterday's? The list goes on almost endlessly.

What's going on here? How can we have good workmanship and good materials and still have problems? What if we do the wrong thing with good materials and good workmanship? Do we still have quality?

Introduction

The problem is not workmanship or materials; the problem is understanding. The pieces must be put together correctly. In order to do so, we must understand how homes work. Homes today work differently than they did in the past. The old solutions and old understandings don't apply.

Homes of yesterday were uncomfortable—too cold in the winter, too hot in the summer—but they usually stood the test of time. Houses of today are exceptionally comfortable but frequently experience serious problems long before the initial mortgage is fully paid. Can there be any connection between comfort and durability? The answer is, "Yes". In a strange way, what we do to homes to make them more comfortable has in fact made them less durable.

In the last fifty years there have been three important changes to the way we build homes:

- The introduction of thermal insulation
- The development of tighter building enclosures
- The advent of forced air heating and cooling systems

Each of these changes has made homes more comfortable, but also has made the same houses less durable.

Thermal Insulation

Thermal insulation was added to wall cavities and ceilings to keep the heat in in the winter, keep the heat out in the summer and make a home more comfortable. However, by keeping the heat in or the heat out, the insulation kept the heat out of the wall cavities and ceilings themselves. In doing so, the ability of these assemblies to dry when they get wet from either interior or exterior sources was reduced. How do you dry something? You heat it. No heat flow, no drying. The addition of thermal insulation increased the "wetting potential" of building enclosures while reducing their "drying potential."

Tighter Building Enclosures and New Materials

Homes built today are much tighter than the homes of yesterday. We use plywood and gypsum board in place of board sheathing and plaster. We platform frame instead of balloon frame. We use factory-made windows instead of site glazing our windows. Building papers come in 10 foot wide rolls instead of 3 foot wide rolls. We put more caulk and glue on our houses than ever before, and we can buy material that actually sticks and holds. The results are fewer holes and a lower air change. The lower the air change, the less the dilution of interior pollutants such as moisture (from people, soil and appliances), formaldehyde (from particle board, insulation, furniture and kitchen and bathroom cabinets), volatile organic compounds (from carpets, paints, cleaners and adhesives), radon (from basements, slabs, crawl spaces and water supplies) and carbon dioxide (from people).

This trend to lower air change occurred simultaneously with the introduction of hundreds of thousands of new chemical compounds, materials and products that were developed to satisfy the growing consumer demand for household goods and furnishings. Interior pollutant sources have increased while the dilution of these pollutants has decreased. As a result, indoor air pollutant concentrations have increased.

Additionally, chimneys don't work well in tight homes. In tight homes, other exhaust fans compete with chimneys and flues for available air. The chimneys and flues typically lose in the competition for available air, resulting in spillage of combustion products, and backdrafting of furnaces and fireplaces.

As air change goes down, interior moisture levels rise causing condensation problems on windows, mold on walls, dust mites in carpets and decay in wall cavities and attic spaces. Yet even though interior moisture levels are rising, builders continue to install central humidifiers rather than installing dilution ventilation or dehumidification.

Traditional chimneys in many new homes have been replaced with power vented, sealed combustion furnaces. Many new homes have no chimneys or flues and rely on heat pumps or electric heating. Traditional chimneys ("active chimneys") acted as exhaust fans. They extracted great quantities of air from the conditioned space that resulted in frequent air changes and the subsequent dilution of interior pollutants. Eliminating the "chimney fan" has led to an increase in interior pollutant levels such as moisture.

Active chimneys also tended to depressurize conditioned spaces during heating periods. Depressurization led to a reduced wetting of building assemblies from interior air-transported moisture and therefore a more forgiving building envelope.

Heating and Cooling Systems

Today, forced air systems (heating and air conditioning) move large quantities of air within building enclosures of increasing tightness. The tighter the building enclosure, the easier it is to pressurize or depressurize. This has led to serious health, safety, durability, and operating cost issues.

Supply duct systems are typically more extensive than return duct systems. There are usually supply registers in each room, with common returns. Pressurization of rooms and depressurization of common areas is created by the combination of more extensive supply systems, leaky returns combined with interior door closure.

When typically leaky supply ducts are run outside the building envelope in vented attic roof and crawl spaces, depressurization of the building enclosure occurs. Depressurization can cause infiltration of radon, moisture, pesticides and soil gas into foundations as well as probable spillage and backdrafting of combustion appliances and potential flame roll-out resulting in fire.

Leaky return ducts and chases connected to exterior spaces can lead to pressurization of the building enclosure. Pressurization can lead to the exfiltration of warm moisture-laden air into wall and roof cavities that are at lower drying potentials because of higher levels of insulation.

Integration

The three important changes in the way we build homes today interact with each other. This is further complicated by the effects of climate and occupant lifestyle. The interrelationship of all of these factors has led to major warranty problems that include health, safety, durability, comfort and affordability concerns. Problems are occurring despite the use of good materials and good workmanship.

We cannot return to constructing drafty building enclosures without thermal insulation, without consumer amenities, and with less efficient heating and air conditioning systems. The marketplace demands sophisticated, high performance buildings operated and maintained intelligently. As such, buildings must be treated as integrated systems that address health, safety, durability, comfort and affordability .

Quality construction consists of more than good materials and more than good workmanship. If you do the wrong thing with good materials and good workmanship, it is still wrong. You must do the right thing with good materials and good workmanship. The purpose of this guide is to promote the use of good materials and good workmanship in a systematic way, so that all the parts work together and promote good performance, durability, comfort, health and safety.

The House System

Functional Relationships

Residential construction is a complex operation including thousands of processes by dozens of industries, bringing together hundreds of components and sub-systems into a house. A house is a complex, interrelated system of people, the building itself and the environment (Figure 1.1).

Figure 1.1
Analytical Model of the House System — Functional Relationships

A house consists of the building envelope, the sub-systems contained within it and the fit and finish. The building envelope is composed of assemblies. Assemblies are composed of elements. Sub-systems are composed of components. The fit and finish is composed of surfaces, appliances, trim, fixtures and the furnishings.

The building envelope, assemblies, elements, sub-systems, components and the fit and finish are all interrelated. A change in an element can change the performance of an assembly, affect the building envelope and subsequently change the characteristics of the house. Similarly, a change in a sub-system can influence the house, an assembly or an element of an assembly.

The house, in turn, interacts with the people who live in the house and with the local environment where the house is located. The functional relationships between the parameters are driven by physical, chemical and biological reactions. The basic factors controlling the physical, chemical and biological reactions are:

- heat flow
- air flow
- moisture flow

Controlling heat flow, air flow and moisture flow will control the interactions among the physical elements of the house, its occupants and the environment.

Building houses is really about the durability of people (health, safety and well being of people), the durability of buildings (the useful service life of a building is typically limited by its durability) and the durability of the planet (the well being of the local and global environment).

The relationships that define the major elements of the residential construction process as well as continuing home operation are represented in Figures 1.2, 1.3, and 1.4.

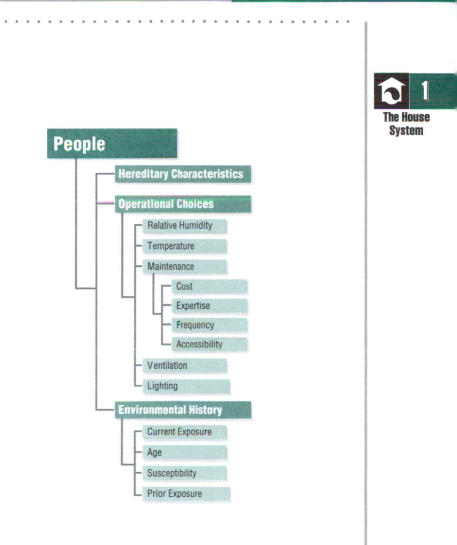

The House System

Figure 1.2
Hierarchical Relationships – People

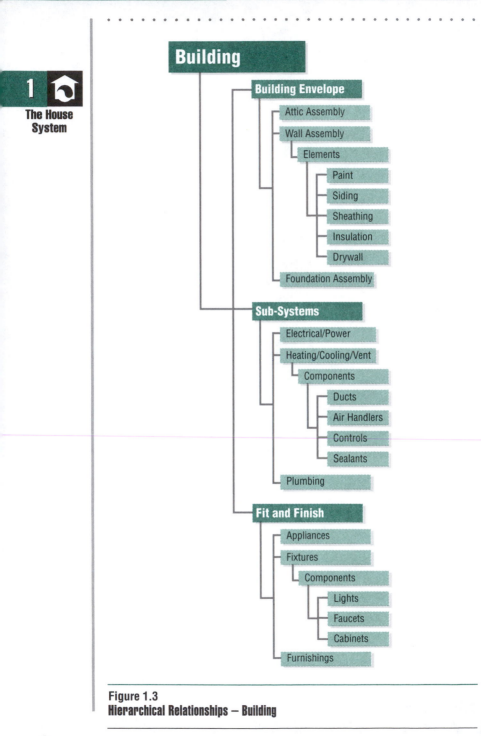

Figure 1.3
Hierarchical Relationships – Building

The House System

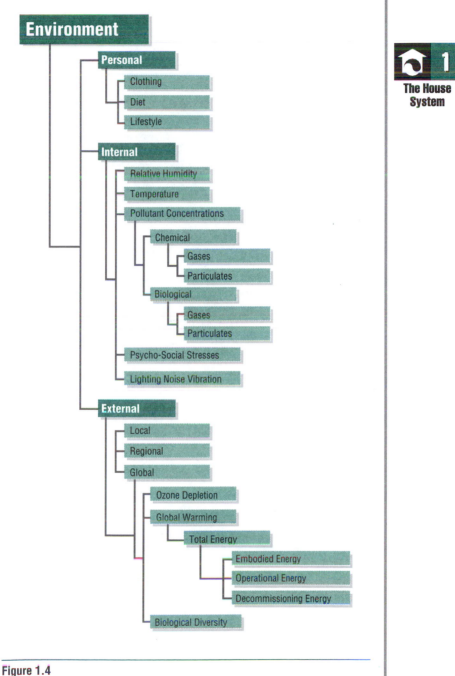

Figure 1.4
Hierarchical Relationships – Environment

1

The House System

Prioritization

The residential construction process should minimize needs for energy, water and materials and satisfy these needs in the least disruptive manner possible (Figure 1.5).

The interior environment, or conditioned space, should be safe, healthy and comfortable. The building should be both durable and affordable in terms of purchase price and operating costs. And the building should be built in a manner that does the least harm to the local environment, including the construction site and the land around it. The impact of home production on the global environment should also be considered. What resources will be used to build and operate the building? Are they renewable? What is the effect on the environment of extracting them?

**Figure 1.5
Minimization of Needs**

The sometimes conflicting needs among people, buildings and the environment should be prioritized. For example, the needs of people should be considered before the needs of a building. The internal environment created by a building should be considered before the planetary environment. Short-term concerns should be considered before long-term concerns (Figure 1.6). But all should be considered.

Applying prioritization to residential construction would identify the immediate health risk from carbon monoxide poisoning as a result of improper installation of combustion appliances as more significant than long-term health concerns from the infiltration of radon gas.

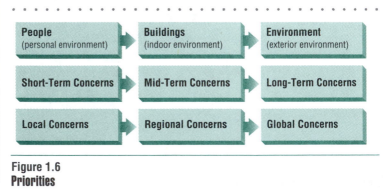

**Figure 1.6
Priorities**

Extending the prioritization process further shows that job-site recycling and reducing construction waste should have precedence over a global concern such as ozone depletion in the upper atmosphere. Finally, ozone depletion — a global short-term risk — should have precedence over global warming — a global long-term risk.

People Priorities

Houses should be safe, healthy, comfortable and affordable. A safe, healthy home is one in which concerns about structural adequacy, fire and smoke spread, security and indoor air quality have been addressed. Comfort involves satisfying people's sensory perception. It implies dealing with thermal comfort, interior relative humidity, odors, natural light, sound and vibrations. Affordability means the designer, the builder and the various subcontractors and suppliers should be able to make a profit, and the occupant, for whom the home is built, should be able to purchase it and afford the operating and maintenance costs (Figure 1.7).

Building Priorities

Houses should be durable and capable of being maintained. The single most important factor affecting durability is deterioration of materials by moisture. Houses should be protected from wetting during construction and operation, and be designed to dry should they get wet.

Homeowners and occupants should be instructed on how to maintain and operate their buildings. Houses should be able to be renewed and renovated as new technologies, materials and products emerge. The house built today will likely be renovated at some point in the future. Houses should be able to be adapted as families and occupancy change. Finally, houses should be designed and constructed with decommissioning at the end of the useful service life in mind. This means taking into consideration how the materials that go into the house will ultimately be disposed of (Figure 1.7).

Environmental Priorities

Houses should be constructed in a manner that reduces construction waste and operated in a manner that reduces occupancy waste. Recycling of construction and operating waste should be encouraged. Use of construction water, domestic water and irrigation water should be minimized. Erosion of soil during site preparation and the construction process should be controlled. Storm water should be infiltrated back into the site.

Activities that contribute to air pollution during construction, such as construction dust, painting and burning of trash, should be minimized. Materials and systems that contribute to ozone depletion by releasing chlorofluorocarbons (CFCs) into the air should be avoided.

Biological diversity of plant and animal species should be protected by using materials from managed forests and managed mineral extraction processes. Finally, the use of energy to operate the building and to make and transport building products (embodied energy) should be minimized to reduce the production of greenhouse gases (e.g., carbon dioxide) that contribute to global warming (Figure 1.7).

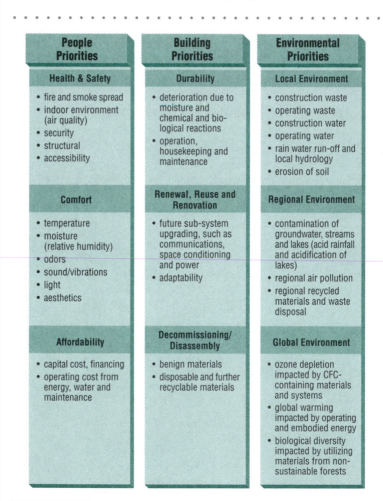

People Priorities	Building Priorities	Environmental Priorities
Health & Safety • fire and smoke spread • indoor environment (air quality) • security • structural • accessibility	**Durability** • deterioration due to moisture and chemical and biological reactions • operation, housekeeping and maintenance	**Local Environment** • construction waste • operating waste • construction water • operating water • rain water run-off and local hydrology • erosion of soil
Comfort • temperature • moisture (relative humidity) • odors • sound/vibrations • light • aesthetics	**Renewal, Reuse and Renovation** • future sub-system upgrading, such as communications, space conditioning and power • adaptability	**Regional Environment** • contamination of groundwater, streams and lakes (acid rainfall and acidification of lakes) • regional air pollution • regional recycled materials and waste disposal
Affordability • capital cost, financing • operating cost from energy, water and maintenance	**Decommissioning/ Disassembly** • benign materials • disposable and further recyclable materials	**Global Environment** • ozone depletion impacted by CFC-containing materials and systems • global warming impacted by operating and embodied energy • biological diversity impacted by utilizing materials from non-sustainable forests

Figure 1.7
People, Building and Environmental Priorities

Home Designer

The designer makes fundamental decisions about the siting, massing, layout and design of the house. The designer is often the general contractor but can be an architect or the home buyer. In many cases, design decisions are shared among the designer, general contractor, and home buyer. The designer must be aware of the limitations of the project budget, the needs of the home buyer and the requirements of the general contractor. The designer must also understand the local climate and the specific limitations of the proposed site. Furthermore, the designer must understand the project time, labor, construction sequence, material characteristics and the process of construction (Figure 2.1).

Site Planning

The ideal site is seemingly never available, and there is never enough money. However, where flexibility exists, sites facing southeast, south or southwest provide the best opportunities for optimizing a building's orientation with respect to daylighting and passive solar gain. Sites sheltered from winter winds and open to summer breezes are warmer in winter and cooler in summer. Bodies of water and areas of vegetation moderate air temperature. Sites shaded by deciduous trees are cooler in summer. Sites that are well drained reduce the stress on drainage systems and water management. Building in a swamp is always more difficult than building on the top of a hill.

The site access, site clearing, excavation, site manipulation and shaping, construction process, site development and landscaping all need to be considered with respect to soil erosion and the existing hydrology. Sculpting the ground to permit a slab-on-grade rather than a walkout basement or elevated crawl space can save thousands of dollars, but may also create severe environmental stress.

2

Home Design

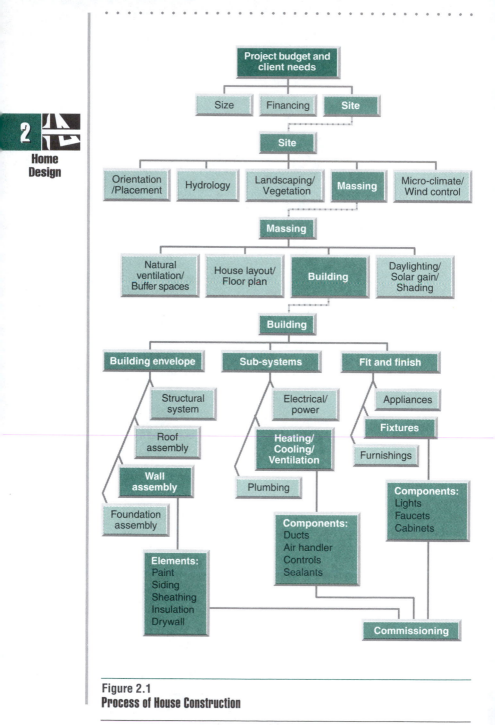

Figure 2.1
Process of House Construction

Landscaping should be used to buffer the house from winter winds, allow winter solar gain and daylighting, and provide summer shading and cooling. Vegetation, walls, fences and other buildings can be used as wind breaks. Wind breaks can be used to channel breezes into buildings and outdoor spaces. Overhead structures can be used to provide shade for outdoor use areas. West and southwest facades, that provide the greatest potential for summer overheating, should be shaded from low-angled sun. Light-colored walls or fences can be used to assist daylighting by reflecting sunlight into north windows. Paving should be minimized and shaded from the sun. Understory vegetation should be cleared and maintained to admit summer breezes.

2

Home Design

Site hydrology and the management of storm water have a major environmental impact. Under natural conditions most rainfall percolates into the ground upon which it falls, whereas almost all the rain that falls on a built-up area contributes to surface run-off. The principle advantage of vegetation is that it controls soil erosion from run-off water. Grass encourages percolation to the water table and is effective vegetation for erosion control. In fact, many grasses work to reduce the impact of pollutants by breaking them down before they reach the water table. Turf grass requires considerably more water than other ground covers. Xeriscaping (vegetation or amenities which require little or no water) should be considered.

An ideal situation would be one in which no increase in run-off occurs as a result of development, so that problems are localized rather than passed on to others. This concept can be promoted by reducing the amount of paving and other impervious surfaces or using porous pavement in order to permit a greater portion of the storm water to seep into the ground. Vegetation can be increased and/or retained in order to maximize the amount of storm water consumed and stored by plants. Storm water can be drained to temporary or permanent storage areas by means of surface collection and low volume underground systems. It's okay to have puddles of water after a rain. Just don't have those puddles right next to the house.

House Layout

The building form and layout are major factors influencing cost and performance. Heat gain and heat loss occur across building surfaces. Maximizing volume while minimizing surface area will increase operating efficiencies. Building forms that are compact are more resource and energy efficient than those that are spread out. In terms of floor plans, locating the most actively used spaces where they will benefit most from daylighting makes the most sense. In cold climates, kitchens, living rooms and family rooms should be located to the south side.

2

Home Design

Buffer spaces should be used both within a building (vestibules) and external to a building (porches, sunspaces, sheltered patios) to temper weather extremes.

Outdoor use areas used primarily during winter months should be located adjacent to the south side of the buildings and southern exposures should remain free from obstructions except with respect to shading from the summer sun. Large paved surfaces should be avoided on windward sides of buildings. Such surfaces, if necessary, should be located to the lee side of use areas with respect to summer breezes so as to minimize summer thermal mass effects.

Open floor plans allow for air flow and ventilation efficiencies, increased daylighting and summer cross ventilation. During heating, heat is more evenly distributed.

Windows should be sized and positioned so as to decrease heat loss during the winter and decrease heat gain in the summer while providing daylighting year round. Moderate amounts of southern glazing provide improved comfort during heating (sun tempering) while providing views without an energy penalty. Substantial south-facing glass can lead to overheating in summer as well as winter and should be avoided. As a minimum, spectrally selective glazing ("southern low-E glass") should be installed throughout (see Appendix V).

In designing for cooling load, west- and southwest-facing windows should be minimized as they cause the most summertime overheating.

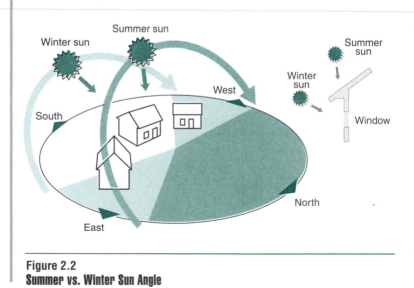

Figure 2.2
Summer vs. Winter Sun Angle

Large areas of glazing overhead (large skylights) should be avoided. Overhangs (Figure 2.2) and vegetation such as deciduous trees should be considered to shade windows during the summertime while leaving windows unshaded in wintertime. Fencing and/or a trellis can also be used to shade glazing, particularly on east and west exposures.

Light-colored walls, floors and ceilings should be used to reflect incoming light deep into a room. A small skylight can be used to get the light to the back of a room. Light ground colors should be avoided in front of south-facing windows to minimize reflected summer heat gain, but should be considered for walls and fences on north exposures to reflect sunlight into north windows for improved daylighting.

Basic Structure and Dimensions

The house layout and massing define the basic structure. The type of foundation system (crawl space, slab or basement), roof system (attic, cathedral ceiling or flat), floor system (joist, truss or slab) and structural system (wood frame, steel frame, concrete or masonry) is selected by the designer based on costs, availability of materials, regional practices and preferences, site conditions and micro-climate, environmental impact and availability of experienced trades.

Plywood, oriented strand board (OSB), and insulating sheathings all come in 4-foot by 8-foot sheets. It makes sense to design out-to-out dimensions on 2-foot increments to reduce sheet good waste and to maximize the efficiency of the structural frame (Figure 2.3).

Roof slopes and overhangs should also be selected and dimensioned to take advantage of 2-foot increments. We often specify 4:12, 6:12 or 8:12 roof pitches when we can just as easily specify a 4.217:12 roof pitch and minimize sheet good waste (Figure 2.3).

Framing members should be spaced on 24-inch centers rather than 16-inch centers. Using this approach, coupled with single plates, stack framing, two stud corners, and elimination of cripples, it is possible to frame with 2x6's less expensively than with 2x4's. The volume (board footage) of lumber is about the same, but there are 30% fewer pieces, so the building goes together faster (less labor/less cost). In this way the wall is thicker and therefore can accommodate more insulation with no additional cost.

Roof framing should line up with wall framing and floor framing. Windows and doors should be located on 2-foot grid increments to eliminate extra studs on the sides of openings.

Fireplaces should not be located on exterior walls, they should be totally contained within the conditioned space. A warm chimney drafts

2 **Home Design**

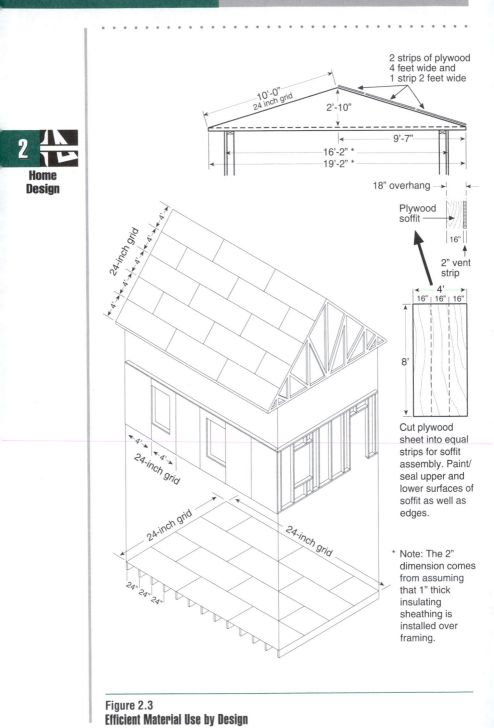

2 strips of plywood
4 feet wide and
1 strip 2 feet wide

10'-0"
24 inch grid

2'-10"

9'-7"

16'-2" *

19'-2" *

18" overhang

Plywood soffit

16"

2" vent strip

4'

16" | 16" | 16"

8'

Cut plywood sheet into equal strips for soffit assembly. Paint/seal upper and lower surfaces of soffit as well as edges.

* Note: The 2" dimension comes from assuming that 1" thick insulating sheathing is installed over framing.

24-inch grid

4' 4' 4'

24-inch grid

4' 4'

24-inch grid

24-inch grid

24" 24" 24"

Figure 2.3
Efficient Material Use by Design

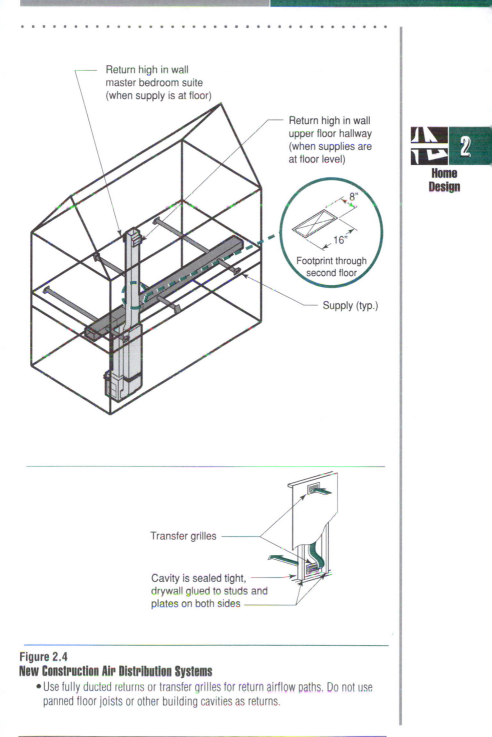

Return high in wall
master bedroom suite
(when supply is at floor)

Return high in wall
upper floor hallway
(when supplies are
at floor level)

8"

16"

Footprint through
second floor

Supply (typ.)

Home
Design

Transfer grilles

Cavity is sealed tight,
drywall glued to studs and
plates on both sides

**Figure 2.4
New Construction Air Distribution Systems**

- Use fully ducted returns or transfer grilles for return airflow paths. Do not use
  panned floor joists or other building cavities as returns.

much better than a cold one. In addition, masonry chimneys that are located within the conditioned space have thermal mass that store the heat created by a fire and continue to warm the space even after the fire has gone out.

2

Home Design

Mechanical equipment, ductwork and plumbing must not be located in exterior walls, vented attics or vented crawl spaces. All air distribution systems must be located within the conditioned space. Interior chases must be considered and provided during the schematic design phase to make it easy to install services (Figure 2.4). A designer must think about stair locations and openings while allowing ductwork and plumbing to get by them as it goes from one side of the house to the other.

Complaints on how plumbers and HVAC installers butcher framing during rough-in are often heard. However, if the designer does not leave them space, or make it easy for them to install equipment, piping and ductwork, it is often the only way to install the services. The designer almost never talks to the plumber or HVAC contractor during the design phase about locating soil stacks, ductwork layouts and equipment locations. It's time we start the dialog, or stop complaining about cut floor framing, having to pad-out walls to hide plumbing and ducts and paying for tortured ductwork layouts that don't deliver any air.

Building Envelope

The building envelope must:

- hold the building up
- keep the rainwater out
- keep the groundwater out
- keep the wind out
- keep the water vapor out
- keep the soil gas out
- let the water and water vapor out if it gets in
- keep the heat in during the winter
- keep the heat out during the summer
- keep the noise out

The designer has to choose materials, equipment and systems to make all this work. Will exterior sheathings be permeable or impermeable? Will building papers or housewraps be used? What type of air flow retarder systems will be used? What will the thermal resistance of the building envelope be?

A key concept should be used at this point in the design approach — the concept of "break points." A break point denotes the situation where an increase in cost in one area is balanced by a reduction in cost in another area.

For example, increasing the thermal resistance of the building envelope by using insulating sheathing and high performance glazing will result in an increased cost. However, the heating and cooling system can be made much smaller with a resulting decrease in the size and cost of ductwork and equipment. Open-web floor trusses cost more than floor joists. However, it takes much less time, effort and money to install plumbing, electrical work and ductwork within floor trusses than within floor joists. A 2x6 costs more than a 2x4, but if you use far fewer 2x6s than 2x4s, the building frame can be put up faster and, therefore, less expensively.

Home Design

Figure 2.5
Material Selection

Material Selection

There are only a few inherently bad materials, but there are many bad ways to use materials. The use of a material should be put into the context of a system. In general, the system is more important than the material. Once the system is selected, you must determine if the material can perform its intended function as part of that system. What is the risk of using the material to the occupants, to the building and to the local and global environment when used in that system (Figure 2.5)?

Extending the argument further, there are no truly benign materials, only degrees of impact. Nothing is completely risk free. However, risk can be managed. There may be no alternative to a particular toxic ma-

2

Home Design

terial in a specific system, but the use of that material may pose little risk when used properly and provide significant benefits to that system. For example, bituminous dampproofing is a toxic material, but if it is installed on the exterior of a concrete foundation wall there is little risk to the occupants, but there are substantial moisture control benefits to the foundation assembly.

The risk to occupants from a particular synthetic or natural agent in a building product, system or assembly is generally low if that agent is not inhaled or touched. In any case, building products and materials that do not off-gas are preferable to those that do. Less toxic alternatives should be used in place of more toxic materials (Figure 2.6). Remember that these material choices need to be placed in the context of

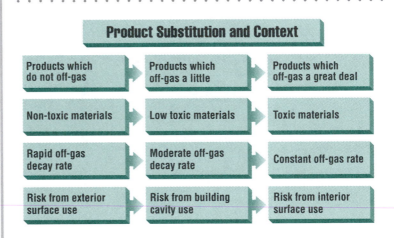

Figure 2.6
Product Substitution and Context

the system or assembly of which it is a part. Is the toxic material being used in roofing? If so, it may pose little hazard to the occupants.

In addition to the specific concerns about material and product use on the interior environment of a building and the occupants, are the concerns relating to the local and global environment. Is it more appropriate to use a recycled, refurbished or remanufactured product or material in place of a new material or product? Is the new material or product obtained or manufactured in a non-disruptive or the least-disruptive manner to the environment? Can the cost of a product justify its use? How far was the material transported? How much energy was used to make it?

Sub-Systems

All buildings require controlled mechanical ventilation. Building intentionally leaky buildings and installing operable windows does not provide sufficient fresh air in a consistent manner. Building envelopes must be "built tight and then ventilated right." Why? Because before you can control air you must enclose it. Once you eliminate big holes, it becomes easy to control air exchange between the inside and the outside. Controlled mechanical ventilation can be provided in many forms (See Chapter 6), but what is important at this stage is that the designer recognize that a controlled mechanical ventilation system is necessary and that controlled mechanical ventilation works best in a tight building envelope. You can't control anything in a leaky building.

Home Design

Selecting a fuel source is usually based on availability, regional practices and customer preference. Electric heat pumps and natural gas are the most common choices for space and domestic hot water heating. If gas heating or a gas water heater is selected, the appliances must be power vented, sealed combustion or installed external to the conditioned space (i.e. gas water heaters installed in a garage). Gas appliances should not interact aerodynamically with the building (i.e. be affected by interior air pressures or other air consuming devices). If a gas cooktop or gas oven is installed, it must be installed in combination with a kitchen range hood directly ducted to the exterior (exhaust fan). Unvented gas fireplaces should never be installed.

Air change through mechanical ventilation can be used during heating periods to control interior moisture levels. Dehumidification through the use of mechanical cooling (air conditioning) can be used during cooling periods to control interior moisture levels.

Air conditioners, furnaces, air handlers and ductwork should be located within the conditioned space and provide easy access to accommodate servicing, filter replacement, cooling coil and drain pan cleaning, future upgrading or replacement as technology improves. Hostile locations (extreme temperatures and moisture levels) such as vented attics and unconditioned (vented) crawl spaces should be avoided.

Appliances

The designer is responsible for the selection of appliances, typically with home owner input. If home owners assume the responsibility of selecting some or all of the appliances, it is the responsibility of the designer to provide the necessary information to the home owner so that an informed decision can be made within the context of the house system.

Appliances should be selected and installed in such a manner that they do not adversely impact the building envelope or building sub-systems. For example, gas cooktops and ovens should provide for their own exhaust of combustion products by installing vented range hoods or exhaust fans. They also need to be installed in a manner which prevents excessive depressurization (i.e. make-up air for indoor barbecues).

2

Home Design

Fireplaces and wood stoves should be considered appliances. They should be provided with their own air supply independent of the other air requirements of the building enclosure. The location of combustion appliances should take into consideration air pressure differentials that may occur due to the stack effect or competition for air from other combustion appliances. Indoor barbecues should be considered similarly. Air supply, air pressure differentials and combustion product venting need to be addressed.

Energy consumption should be a prime consideration in the selection of refrigerators, freezers, light fixtures, washers and dryers. The yellow D.O.E. Energy Guide label should be consulted. Dryers should be vented directly to the exterior and should not adversely affect the air pressure dynamics of the building enclosure when they are operating.

Commissioning

The home designer must ensure that commissioning of the home occurs so that the home functions as intended by the design. Commissioning allows problems to be spotted through testing and then immediately remedied ("fix and tune"). The commissioning can be done by the designer, the contractor or some other competent person. At minimum, commissioning should include:

- testing of the building envelope leakage area
- testing of the leakage of duct systems
- testing of the áir pressure relationships under all operating conditions
- testing for proper venting of all combustion appliances under all operating conditions
- testing of the carbon monoxide output of all combustion appliances (gas oven, range, water heater, gas fireplaces)
- confirmation of airflow and refrigerant charge in HVAC systems

As part of the commissioning process, the home owner or occupants should be educated and informed as to the correct operation, maintenance and housekeeping requirements of the building. What are appropriate temperature, relative humidity and ventilation ranges for the building? How do owners or occupants identify a system failure as distinct from the improper operation of a system? How do owners or occupants monitor the building conditions (temperature sensors, humidity sensors, ventilation sensors)?

There is information relating to air leakage testing of building envelopes leakage areas, testing of the air leakage of duct systems, air pressure differentials and combustion safety in Appendix VI.

Home Design

General Contractor

General Contractor

All the general contractor has to do is construct the building on time and under budget using imperfect materials and imperfect trades, under less than ideal conditions. The client even expects the building to work. There never seems to be enough time or money, but the job still has to get done. In this line of work, Murphy is an optimist.

In the old days, all you needed was good workmanship and good materials. Getting good workmanship today is hard enough given the state of the trades, but it is not enough. Good workmanship cannot compensate for bad design. A general contractor cannot just follow the plans. Plans rarely provide enough information to get the job done, and many times the plans are wrong, in which case, the general contractor has to catch the mistakes. When the plans don't provide enough information, the general contractor has to fill in the gaps. "Not my fault, I was just following the plans," or "Nobody told me I had to do it that way," doesn't work anymore. It's not fair, nor is it right, but often the way it is. In order to protect himself or herself, the general contractor has to know everything about everything. Easy, right?

Concerns

Everything is a concern to the general contractor, but of all the concerns, there is one that towers above the rest. Getting the right information to people when they need it is more important than anything else. Getting the materials and equipment to the people when they need it comes next. General contracting is all about the flow of information, materials and equipment to the right people at the right time.

Most mistakes happen because of a lack of information, followed by a lack of attention. Attention follows information. Understanding comes from having the right information. Once there is understanding, problems can be caught early and corrected by the people actually doing the work. It helps the general contractor to employ workers who understand how their jobs fit into the entire process.

Unfortunately, nowadays it is difficult to find fully trained and knowledgeable trades people. It seems that trades people don't take courses or classes anymore and that apprenticeship programs have fallen by the wayside. The depth of experience in the trades is often missing and the workforce is very transient. Even when there has been training, it is often out of date or just plain wrong. In order to get the job done right, the general contractor has to train the trades himself, everyday, day in and day out.

General Contractor

The most effective training approach is on-site training, beginning with the first day of work. This should occur during a one-hour period prior to commencing work. "This is what I want and why." "This is how I think you should do it, but I'm open to any suggestions on how to do it simpler." "I'm available all day to answer questions."

Training is always easier if you have the right tools. Simple training tools, like visual aids posted on site for easy reference and use, work well. Posters illustrating key details can be developed for framers, electricians, insulators and drywallers.

Detailed framing drawings can be created that illustrate the location of each stud and framing member (Figure 3.1). These drawings can significantly reduce construction time as well as potential confusion. Supervisory time can also be reduced.

The better, the quicker and the simpler the training, the better the subcontractors' prices the next time a job is bid.

The detailed framing drawings and the posters developed for the framers, electricians, insulators and drywallers, coupled with checklists and addresses of suppliers, become key elements in developing "competitive pricing" during bid negotiations.

No one likes surprises. The better the information and the better the training, the fewer the surprises. When there are no surprises, you get good prices. Now we're talking.

Materials on the Building Site

Materials should be delivered to the site just before they are to be used. It they are not there, they don't get stolen, rained on, beat-up or destroyed. Of course if materials are not there when you need them, you also have big trouble. "Just-in-time delivery" is more than a slogan, it should be a way of life.

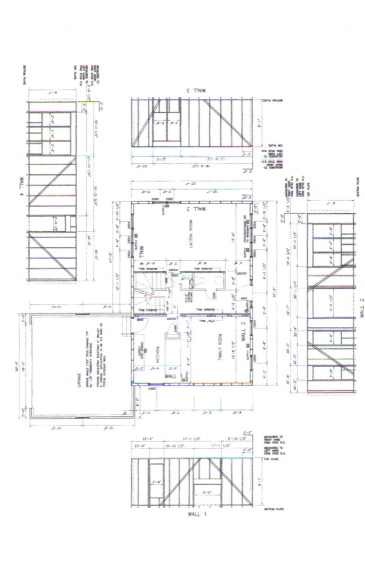

Figure 3.1
Framing Diagram

• Studs, blocking and cross bracing shown on drawings

If you have to store materials at the job site, dumping them on the ground in a pond of water in everyone's way is not generally a good idea. Material should be placed on skids raised off the ground and covered with material to shed water. Material should be placed in a convenient location. Moving material more than once is bad planning. A transport trailer makes a good storage shed if you can keep someone from hooking up their truck and pinching it.

Don't get more than you need. If you get more than you need, it's never left over at the end of the job. It somehow disappears. On job sites the law of conservation of mass does not hold. Here's where detailed drawings and detailed take-offs pay for themselves.

3

General Contractor

Sites should be neat. Sites should be orderly. Sites should be clean. Clean, neat, orderly sites are safe sites and productive sites. He who makes a mess, cleans the mess. Those of us who were kids once, learned to clean up our toys when we were done. A job site should run by the same rules.

"Drying-in" the building as quick as possible is a good idea. Roofing felt is relatively inexpensive and can be used to cover everything. Remember that wood loves water. Gypsum board loves water. Mold loves wood and gypsum board when it is wet. Mold is bad.

Materials that come to the job site "flat" and that need to be "flat" when installed should be stored "flat." Roofing shingles should be stored so that the bundles can lie flat without bending. Windows and sliding doors like to be "square." Store them "square." Remember that big piles of building materials can be real heavy. Heavy loads, concentrated on one spot for any appreciable time can be a real problem. Distribute drywall and other heavy loads. Be smart. Think.

Foundations

The three foundation approaches common to residential construction are crawl spaces, slabs and basements. Each can be built with concrete, masonry or wood. Each can be insulated on the inside, on the outside or from below. However, they all have to:

- hold the building up
- keep the groundwater out
- keep the soil gas out
- keep the water vapor out
- let the water and water vapor out if it gets inside
- keep the heat in during the winter

Concerns

Concrete cracks. Concrete has always cracked. Concrete will always crack. Reinforcing concrete will not prevent it from cracking. It is not possible to build a crack-free concrete slab or foundation wall. However, it is possible to control the cracking process by deliberately cracking the concrete (Figure 4.1). These deliberate cracks are called control joints. Cracks do not naturally occur in straight lines, nor do they happen in predictable places; but control joints cause concrete to crack along straight lines and in predictable locations where builders can better deal with them. Home owners get annoyed at cracks in concrete, but home owners don't have a problem with control joints. Cracks are bad. Control joints are good.

Concrete also shrinks, creeps and moves. This is also true for masonry and brick, only more so. Masonry and brick swell when they get wet. Concrete can shrink while brick is swelling. Concrete moves. Masonry moves. Brick moves. Let them move.

When wood gets wet it expands; when it dries it shrinks. Wood is almost always in the process of either getting wet, absorbing moisture or drying. Therefore, wood is almost always moving. Wood has always moved and will always move. Nailing wood, screwing wood and gluing wood will not prevent it from moving. It is not possible to build a wood wall, floor, roof or foundation and not have it move. Let it move.

Mixing wood with concrete, masonry and brick makes building interesting but also makes buildings move. Buildings will always move. It is not possible to build buildings that do not move. Let them move.

Since buildings move, it is not possible to build one without holes. Builders can reduce the number of holes; builders can control the size of holes. Builders can control the types of holes. But make no mistake about it, there will be holes. The trick is to keep the water out, even though you have holes. Fortunately, water is lazy. Water will always choose the easiest path to travel. If you provide an easy path for water to travel to a foundation drain, it will follow that path rather than a path through a foundation wall, even if the foundation wall has holes. And we know the foundation wall will have holes despite our best efforts.

4

Foundations

Water Managed Foundations

Water managed foundation systems rely on two fundamental principles (Figure 4.2):

- keep rainwater away from the foundation wall perimeter
- drain groundwater with sub-grade perimeter footing drains before it gets to the foundation wall

Water managed foundation systems are different from waterproofing systems. Waterproofing relies on creating a watertight barrier without holes. It can't be done. Even boats need pumps. Water managed foundation systems prevent the buildup of water against foundation walls, thereby eliminating hydrostatic pressure. No pressure, no force to push water through a hole. Remember, we know the foundation wall will have holes.

Mixing control joints with water management is a fundamental requirement for functional foundation systems that provide an extended useful service life.

Dampproofing should not be confused with waterproofing. Dampproofing protects foundation materials from absorbing ground moisture by capillarity. Dampproofing is not intended to resist groundwater forces (hydrostatic pressure). If water management is used, waterproofing is not necessary. However, control of capillary water is still required (dampproofing). Dampproofing is typically provided by coating the exterior of a concrete foundation wall with a tar or bituminous paint or coating.

Draining groundwater away from foundation wall perimeters is typically done with free-draining backfill such as sand, drainage boards or exterior foundation insulations with drainage properties.

Foundations

Soil Gas

Keeping soil gas (radon, water vapor, herbicides, termiticides, methane, etc.) out of foundations cannot be done by building hole-free foundations because hole-free foundations cannot be built. Soil gas moves through holes due to a pressure difference. Since we cannot eliminate the holes, the best thing we can do is control the pressure.

The granular drainage pad located under concrete slabs can be integrated into a sub slab ventilation system to control soil gas migration by creating a zone of negative pressure under the slab. A vent pipe connects the sub slab gravel layer to the exterior through the roof (Figures 4.3, 4.4 and Figure 4.7). An exhaust fan can be added later, if necessary (Figure 4.6).

Moisture

Controlling water vapor in foundations relies first on keeping it out, and second, on letting it out when it gets in. Make no mistake, it will get in. The issue is complicated by the use of concrete and masonry because there are thousands of pounds of water stored in freshly cast concrete and freshly laid masonry to begin with. This moisture of construction has to dry to somewhere, and it usually (but not always) dries to the inside.

Foundations

For example, we put coarse gravel (no fines) and a polyethylene vapor diffusion retarder under a concrete slab to keep the water vapor and water in the ground from getting into the slab from underneath. The gravel and polyethylene do nothing for the water already in the slab. This water can only dry into the building. Installing flooring, carpets or tile over this concrete before it has dried sufficiently and in a manner that does not permit drying, is a common mistake that leads to mold, buckled flooring and lifted tile.

Similarly, we install dampproofing on the exterior of concrete foundation walls and provide a water managed foundation system to keep water vapor and water in the ground from getting into the foundation from the exterior. Again, this does nothing for the water already in the foundation wall. When we then install interior insulation and finishes on the interior of a foundation wall in a manner that does not permit drying to the interior, mold will grow.

Foundation wall and slab assemblies must be constructed so that they resist water vapor and water from getting in them, but they also must be constructed so that it is easy for water vapor to get out when it gets in or if the assembly was built wet to begin with (as they typically are).

Insulating under a concrete slab with vapor permeable or semi-vapor permeable rigid insulation will cause the concrete slab to dry into the ground as well as into the building. By insulating under the slab, the slab becomes much warmer than the ground. Water vapor flows from warm to cold. If the under slab insulation is not a major vapor diffusion retarder, the slab will be able to dry into the ground, even if the ground is saturated. A polyethylene vapor diffusion retarder under a concrete slab is unnecessary when under slab insulation is used. You should use either a polyethylene vapor diffusion retarder or under slab insulation. It is not necessary to use both. However, should you choose to use both, the polyethylene must be in contact with the bottom of the slab to avoid the creation of a reservoir of standing water.

4

Foundations

This approach can also be applied to foundation walls that are insulated on the exterior. By warming the foundation walls relative to the ground, the moisture moves outwards into the ground. Again, the exterior insulation must not be a major vapor diffusion retarder, and dampproofing cannot be installed. The exterior insulation used must also be a capillary break and provide drainage. Only rigid fiberglass, rock and slag wool insulation have these two properties. All other rigid insulations used on the exterior of foundation walls should be used with dampproofing.

Drying a foundation wall assembly or floor slab after it is insulated and after surface finishes have been installed should only be done using diffusion ("letting them breathe"), not air flow ("ventilation"). Allowing interior air (that is usually full of moisture, especially in the humid summer months) to touch cold foundation surfaces will cause condensation and wetting, rather than the desired drying. It is important that interior insulation assemblies and finishes be constructed as airtight as possible but vapor permeable. This will prevent interior moisture-laden air from accessing cold surfaces during both the winter and summer and still allow the assemblies to dry. It is extremely important not to have a vapor diffusion retarder on the interior of interior insulation assemblies in order to permit drying to the interior.

Crawl Spaces

Constructing vented crawl spaces is a bad idea for reasons similar to those explained above. Venting a crawl space with exterior, humid air during summer months leads to the wetting of crawl space assemblies, rather than drying, since crawl space surfaces will be cooler than the outside air. Crawl spaces should be constructed like mini-basements. They should be heated during the winter and cooled during the summer. They should be sealed and enclosed like any other conditioned space, such as a bedroom or a living room. That means that they should have a supply HVAC system outlet. A return HVAC system duct or grille is not typically recommended — the "leakiness" of the floor assembly will provide the return air path. However, a return duct, dampered to prevent crawl space depressurization (via excessive return air flow) is acceptable. Alternatively, a floor grille acting as a "transfer" grille can be installed. It should be noted that installation of a "transfer" grille between the crawl space and the house may be in conflict with some codes unless crawl space services are also fire-rated. The crawl space may be considered a plenum with such an approach. Accidental "active" depressurization of the conditioned crawl spaces via a return duct or return duct leakage that creates recirculation into the house should be avoided. This should not be confused with continuous, active depressurization of crawl spaces via an exhaust fan vented to the exterior that is a proven soil gas control strategy (Figure 4.5).

If, for unavoidable reasons, a crawl space must be vented, the crawl space floor assembly should be constructed like any exterior wall assembly with impermeable insulating sheathing, only with the assembly lying flat—horizontal (Figure 4.21).

Polyethylene Under Slabs

A sand layer is sometimes installed over a polyethylene vapor diffusion retarder located under a concrete slab. It is thought by some that the sand layer will protect the polyethylene from damage and act as a receptor for excess mix water in the concrete slab when the concrete is cast. This is an extremely bad idea because a reservoir of standing water can be created. If groundwater rises sufficiently to contact the underside of the polyethylene, water will enter the sand layer and be held in the sand layer by capillary forces, even after the groundwater level drops. Since the polyethylene is between the water-soaked sand and the ground, the only way for the water to get out is up into the building through the concrete slab by diffusion. The wetting of the sand by groundwater can take only minutes, but the drying out may take a decade.

The polyethylene under a concrete slab can function as an effective vapor diffusion retarder even if it has holes. It does not need to be protected. The best way to deal with excess mix water is to not have any. Use low water-to-cement ratio concrete with an accelerator or a superplasticizer. It is faster and easier and, therefore, less expensive. The concrete costs a little more, but the labor is much less.

Carpets

Installing carpets on cold, damp concrete floor slabs can lead to serious allergic reactions and other health-related consequences. It is not recommended that carpets be installed on basement concrete slabs unless the carpets can be kept dry and warm. In practice, this is not possible

unless basement floor slab assemblies are insulated and basement areas
are conditioned. Installing carpets on concrete slab foundations located
at grade typically does not pose a risk if the carpet and associated carpet pad are vapor permeable. Slabs on grade are typically warmer and
much dryer than basement slabs.

Insects and Termites

There is no good way of dealing with termites. Borate-treated wood
framing, cavity insulation (cellulose) and rigid foams are a promising
approach, but long-term performance has yet to be demonstrated. Using a protective membrane or a stainless steel mesh with a polymer cement slurry as a termite barrier coupled with soil treatment seems to
work on the few projects that have used the approach. However, there
is no universal consensus on this matter as no long term performance
information is available. The protective membranes used have been adhesive-backed roll roofing, waterproofing membranes or ice dam protection membranes. Urethane based sealants are also used to seal the
gaps between basement floor slabs and perimeter foundation walls.
Their long term effectiveness in controlling termite entry is also un-
known. Where protective membranes are exposed to sunlight they must
be resistant to ultraviolet (UV) radiation or protected from UV exposure by using aluminum sheet stock or other alternative materials. In
some foundation details, it may be possible to provide a "viewing
strip" to allow for inspection of possible termite pathways.

4

Foundations

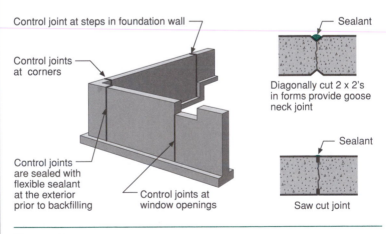

Figure 4.1
Control Joints in Concrete Foundation Walls
- Control joints should be within 10' of corners
- Control joint spacing should be 20' maximum

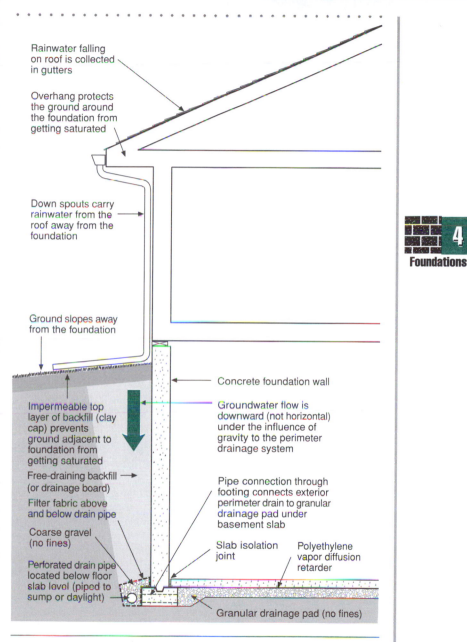

Rainwater falling on roof is collected in gutters

Overhang protects the ground around the foundation from getting saturated

Down spouts carry rainwater from the roof away from the foundation

Ground slopes away from the foundation

Impermeable top layer of backfill (clay cap) prevents ground adjacent to foundation from getting saturated

Free-draining backfill (or drainage board)

Filter fabric above and below drain pipe

Coarse gravel (no fines)

Perforated drain pipe located below floor slab level (piped to sump or daylight)

Concrete foundation wall

Groundwater flow is downward (not horizontal) under the influence of gravity to the perimeter drainage system

Pipe connection through footing connects exterior perimeter drain to granular drainage pad under basement slab

Slab isolation joint

Polyethylene vapor diffusion retarder

Granular drainage pad (no fines)

4

Foundations

Figure 4.2
Water Managed Foundations

- Keep rainwater away from the foundation wall perimeter
- Drain groundwater away in sub-grade perimeter footing drains before it gets to the foundation wall

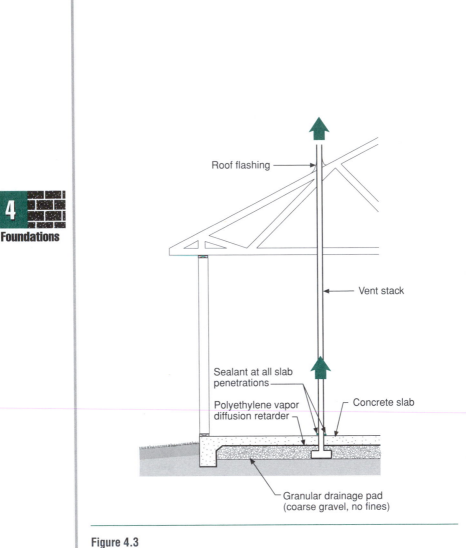

Roof flashing

Vent stack

Sealant at all slab penetrations

Polyethylene vapor diffusion retarder

Concrete slab

Granular drainage pad (coarse gravel, no fines)

Figure 4.3
Soil Gas Ventilation System — Slab Construction

- Granular drainage pad depressurized by active fan located in attic or by passive stack action of warm vent stack located inside heated space
- Communication to all sub-slab areas is required. Where slabs are divided by a thickened section to support a bearing wall, pipe connections through the thickened section will be necessary. Multiple connection points or interconnection piping may be required between multi-level slabs similar to those found in split level homes.

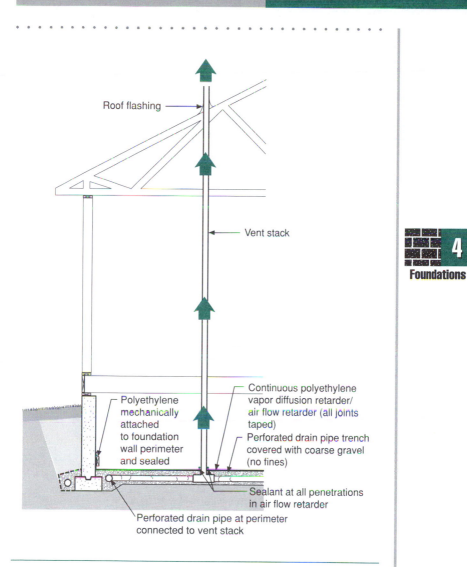

Roof flashing

Vent stack

Polyethylene
mechanically
attached
to foundation
wall perimeter
and sealed

Continuous polyethylene
vapor diffusion retarder/
air flow retarder (all joints
taped)

Perforated drain pipe trench
covered with coarse gravel
(no fines)

Sealant at all penetrations
in air flow retarder

Perforated drain pipe at perimeter
connected to vent stack

4

Foundations

Figure 4.4
Soil Gas Ventilation System — Crawl Space Construction

- Perforated drain pipe in trenches covered with coarse gravel create depressurized zones under air flow retarder due to active fan located in attic or by passive stack action of warm vent stack located inside heated space
- Crawl space is conditioned (heated during the winter, cooled during the summer) by a supply HVAC system duct
- Perforated drain pipe may not be necessary with tightly sealed polyethylene and coarse gravel
- Perimeter trench connected to centrally located vent stack

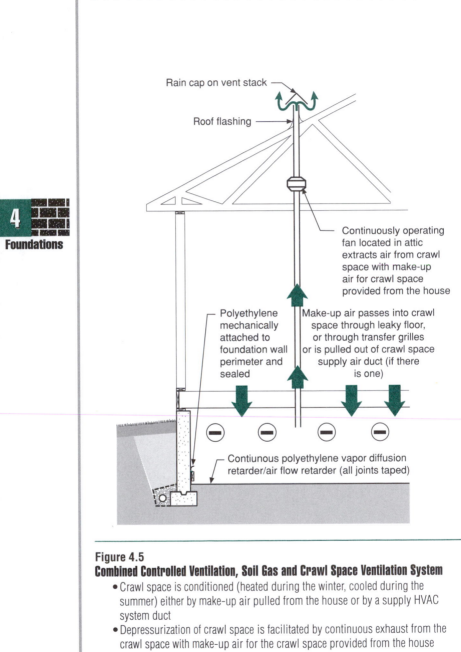

Rain cap on vent stack

Roof flashing

Continuously operating fan located in attic extracts air from crawl space with make-up air for crawl space provided from the house

Polyethylene mechanically attached to foundation wall perimeter and sealed

Make-up air passes into crawl space through leaky floor, or through transfer grilles or is pulled out of crawl space supply air duct (if there is one)

Contiunous polyethylene vapor diffusion retarder/air flow retarder (all joints taped)

Figure 4.5
Combined Controlled Ventilation, Soil Gas and Crawl Space Ventilation System

- Crawl space is conditioned (heated during the winter, cooled during the summer) either by make-up air pulled from the house or by a supply HVAC system duct
- Depressurization of crawl space is facilitated by continuous exhaust from the crawl space with make-up air for the crawl space provided from the house common area
- Crawl space ground cover is tighter than subfloor
- Crawl space ventilation and house ventilation is provided by a single fan

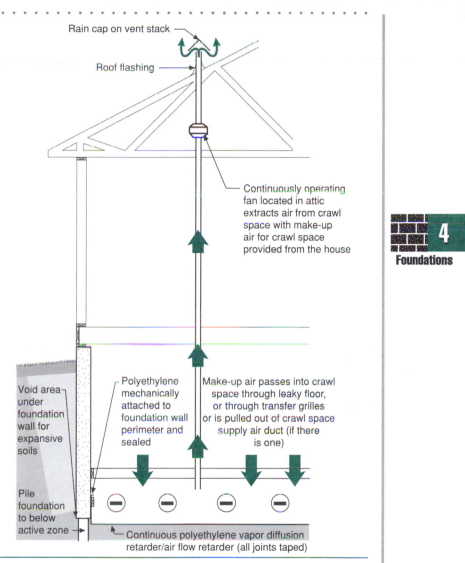

Rain cap on vent stack

Roof flashing

Continuously operating fan located in attic extracts air from crawl space with make-up air for crawl space provided from the house

4

Foundations

Void area under foundation wall for expansive soils

Polyethylene mechanically attached to foundation wall perimeter and sealed

Make-up air passes into crawl space through leaky floor, or through transfer grilles or is pulled out of crawl space supply air duct (if there is one)

Pile foundation to below active zone

Continuous polyethylene vapor diffusion retarder/air flow retarder (all joints taped)

Figure 4.6
Combined Controlled Ventilation, Soil Gas and Crawl Space Ventilation System

- Crawl space is conditioned (heated during the winter, cooled during the summer) either by make-up air pulled from the house or by a supply HVAC system duct
- In regions with expansive soils and suspended wood basement floors over crawl spaces, depressurization of crawl space can be facilitated by continuous exhaust from the crawl space with make-up air for the crawl space provided from the house common area
- Crawl space ground cover is tighter than subfloor
- Crawl space ventilation and house ventilation is provided by a single fan

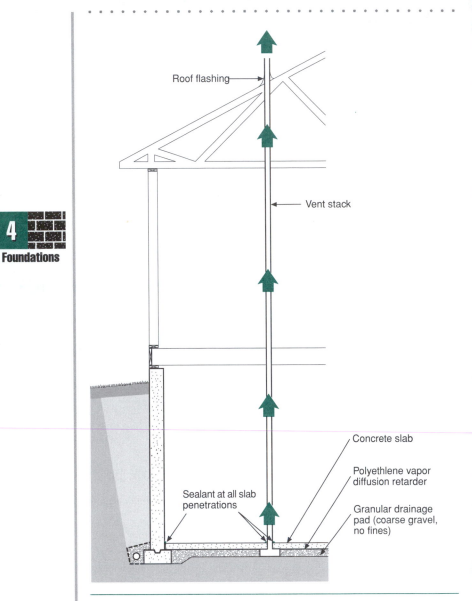

Figure 4.7
Soil Gas Ventilation System — Basement Construction
- Granular drainage pad depressurized by active fan located in attic or by passive stack action of warm vent stack located inside heated space

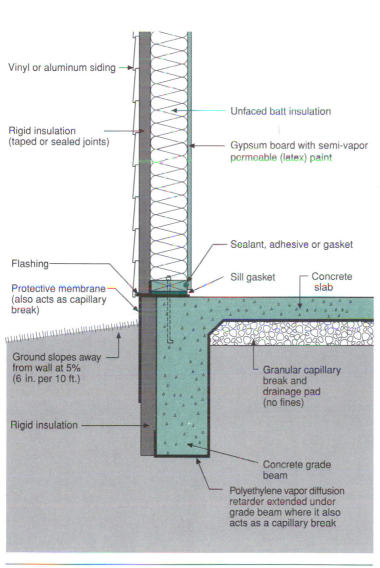

Vinyl or aluminum siding

Unfaced batt insulation

Rigid insulation
(taped or sealed joints)

Gypsum board with semi-vapor
permeable (latex) paint

4
Foundations

Sealant, adhesive or gasket

Flashing

Sill gasket

Concrete
slab

Protective membrane
(also acts as capillary
break)

Ground slopes away
from wall at 5%
(6 in. per 10 ft.)

Granular capillary
break and
drainage pad
(no fines)

Rigid insulation

Concrete grade
beam

Polyethylene vapor diffusion
retarder extended under
grade beam where it also
acts as a capillary break

Figure 4.8
Monolithic Slab — Vinyl or Aluminum Siding
- Protective membrane acts as termite barrier-sealed to slab
- Protective membrane can be adhesive-backed roll roofing or other UV resistant
  materials. Below grade sheet waterproofing or ice-dam protection membranes
  can also be used if protected from UV exposure by using aluminum sheet stock
  or other alternative materials.

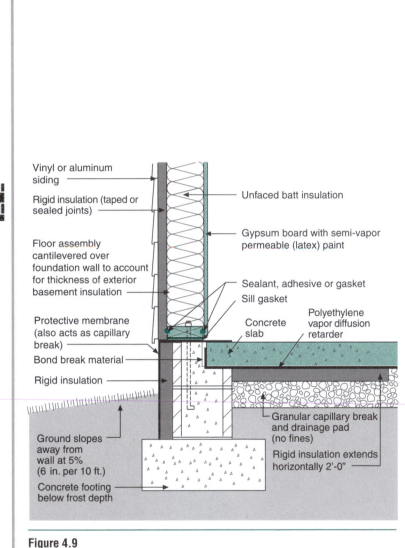

Figure 4.9
Slab with Masonry Perimeter — Vinyl or Aluminum Siding

- Protective membrane acts as termite barrier – sealed to slab
- Polyethylene under slab also acts as capillary break at slab/masonry perimeter
- Protective membrane can be adhesive-backed roll roofing or other UV resistant materials. Below grade sheet waterproofing or ice-dam protection membranes can also be used if protected from UV exposure by using aluminum sheet stock or other alternative materials.
- In locations with low water tables, the interior slab insulation may be omitted due to the lower soil conductivity (heat loss).

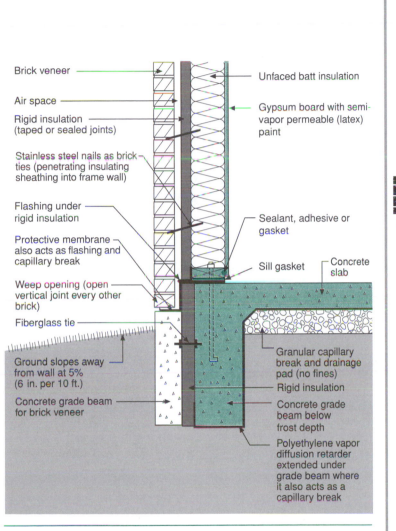

Brick veneer

Air space

Rigid insulation
(taped or sealed joints)

Stainless steel nails as brick
ties (penetrating insulating
sheathing into frame wall)

Flashing under
rigid insulation

Protective membrane
also acts as flashing and
capillary break

Weep opening (open
vertical joint every other
brick)

Fiberglass tie

Ground slopes away
from wall at 5%
(6 in. per 10 ft.)

Concrete grade beam
for brick veneer

Unfaced batt insulation

Gypsum board with semi-
vapor permeable (latex)
paint

Sealant, adhesive or
gasket

Sill gasket

Concrete
slab

Granular capillary
break and drainage
pad (no fines)

Rigid insulation

Concrete grade
beam below
frost depth

Polyethylene vapor
diffusion retarder
extended under
grade beam where
it also acts as a
capillary break

4

Foundations

**Figure 4.10
Monolithic Slab — Brick Veneer**

- Protective membrane acts as termite barrier and acts as flashing at base of brick
  veneer-sealed to slab
- Grade beam for brick veneer cast simultaneously with monolithic slab
- Airspace behind brick veneer can be as small as $3/8$", 1" is typical

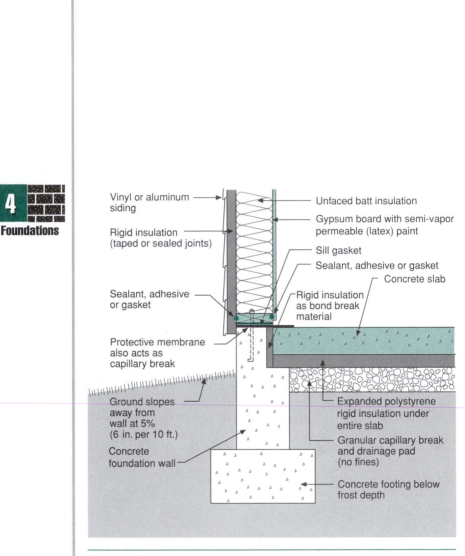

Vinyl or aluminum siding

Unfaced batt insulation

Rigid insulation (taped or sealed joints)

Gypsum board with semi-vapor permeable (latex) paint

Sill gasket

Sealant, adhesive or gasket

Concrete slab

Sealant, adhesive or gasket

Rigid insulation as bond break material

Protective membrane also acts as capillary break

Ground slopes away from wall at 5% (6 in. per 10 ft.)

Concrete foundation wall

Expanded polystyrene rigid insulation under entire slab

Granular capillary break and drainage pad (no fines)

Concrete footing below frost depth

Figure 4.11
Slab with Concrete Perimeter — Vinyl or Aluminum Siding

- Protective membrane acts as termite barrier – sealed to slab
- Rigid insulation on frame wall extends downward below top of concrete foundation wall to shelter horizontal joint
- Floor slab is warm due to sub-slab rigid insulation under entire floor; can dry to the ground (since there is no under slab vapor diffusion retarder, retarder insulation selected is semi-permeable) as well as to the interior, lowest likelihood of mold

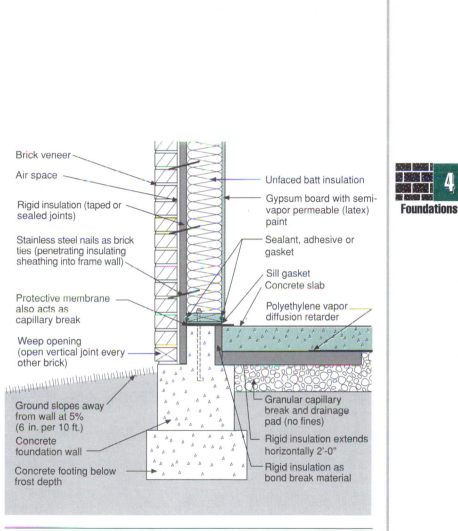

Brick veneer

Air space

Rigid insulation (taped or sealed joints)

Stainless steel nails as brick ties (penetrating insulating sheathing into frame wall)

Protective membrane also acts as capillary break

Weep opening (open vertical joint every other brick)

Ground slopes away from wall at 5% (6 in. per 10 ft.)

Concrete foundation wall

Concrete footing below frost depth

Unfaced batt insulation

Gypsum board with semi-vapor permeable (latex) paint

Sealant, adhesive or gasket

Sill gasket
Concrete slab

Polyethylene vapor diffusion retarder

Granular capillary break and drainage pad (no fines)

Rigid insulation extends horizontally 2'-0"

Rigid insulation as bond break material

4
Foundations

Figure 4.12
Slab with Concrete Perimeter — Brick Veneer

- Protective membrane acts as termite barrier – sealed to slab
- Airspace behind brick veneer can be as small as $^3/_8$", 1" is typical

4

Foundations

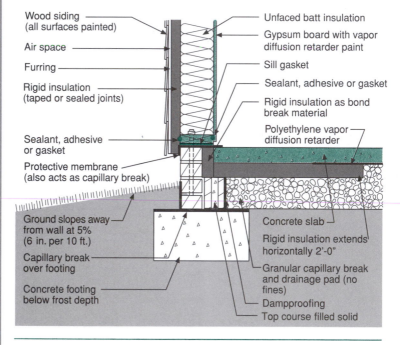

Wood siding
(all surfaces painted)

Air space

Furring

Rigid insulation
(taped or sealed joints)

Sealant, adhesive
or gasket

Protective membrane
(also acts as capillary break)

Ground slopes away
from wall at 5%
(6 in. per 10 ft.)

Capillary break
over footing

Concrete footing
below frost depth

Unfaced batt insulation

Gypsum board with vapor
diffusion retarder paint

Sill gasket

Sealant, adhesive or gasket

Rigid insulation as bond
break material

Polyethylene vapor
diffusion retarder

Concrete slab

Rigid insulation extends
horizontally 2'-0"

Granular capillary break
and drainage pad (no
fines)

Dampproofing

Top course filled solid

Figure 4.13
Slab with Masonry Perimeter — Wood Siding
- Protective membrane acts as termite barrier – sealed to slab

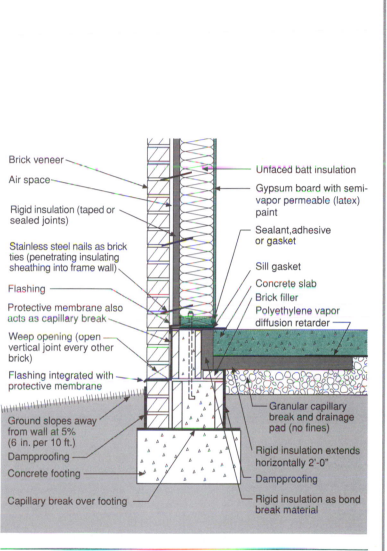

Brick veneer

Air space

Rigid insulation (taped or sealed joints)

Stainless steel nails as brick ties (penetrating insulating sheathing into frame wall)

Flashing

Protective membrane also acts as capillary break

Weep opening (open vertical joint every other brick)

Flashing integrated with protective membrane

Ground slopes away from wall at 5% (6 in. per 10 ft.)

Dampproofing

Concrete footing

Capillary break over footing

Unfaced batt insulation

Gypsum board with semi-vapor permeable (latex) paint

Sealant, adhesive or gasket

Sill gasket

Concrete slab

Brick filler

Polyethylene vapor diffusion retarder

Granular capillary break and drainage pad (no fines)

Rigid insulation extends horizontally 2'-0"

Dampproofing

Rigid insulation as bond break material

4

Foundations

Figure 4.14
Slab with Masonry Perimeter — Brick Veneer

- Protective membrane acts as termite barrier – sealed to slab
- Brick veneer corbelled outwards at base to provide drainage space
- Airspace behind brick veneer can be as small as $^3/_8$", 1" is typical

4

Foundations

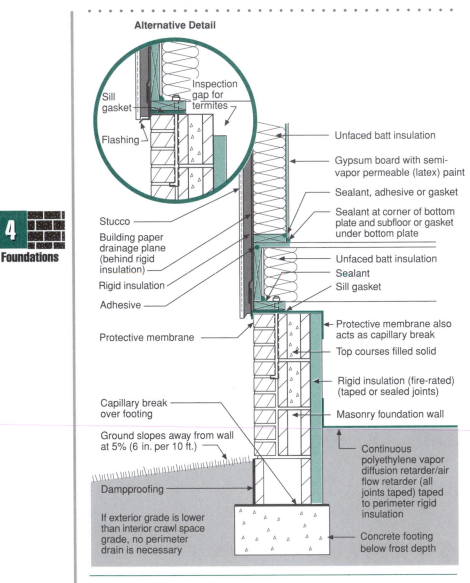

Alternative Detail

Sill gasket

Inspection gap for termites

Flashing

Unfaced batt insulation

Gypsum board with semi-vapor permeable (latex) paint

Sealant, adhesive or gasket

Sealant at corner of bottom plate and subfloor or gasket under bottom plate

Stucco

Building paper drainage plane (behind rigid insulation)

Unfaced batt insulation

Sealant

Sill gasket

Rigid insulation

Adhesive

Protective membrane

Protective membrane also acts as capillary break

Top courses filled solid

Rigid insulation (fire-rated) (taped or sealed joints)

Masonry foundation wall

Capillary break over footing

Ground slopes away from wall at 5% (6 in. per 10 ft.)

Continuous polyethylene vapor diffusion retarder/air flow retarder (all joints taped) taped to perimeter rigid insulation

Dampproofing

If exterior grade is lower than interior crawl space grade, no perimeter drain is necessary

Concrete footing below frost depth

Figure 4.15
Internally Insulated Masonry Crawl Space — Stucco
- Masonry wall cold, can dry to the exterior; low likelihood of mold
- Protective membrane acts as termite barrier
- Rigid insulation must be fire-rated if it is left exposed on the interior
- Building paper installed shingle fashion acts as drainage plane located behind rigid insulation

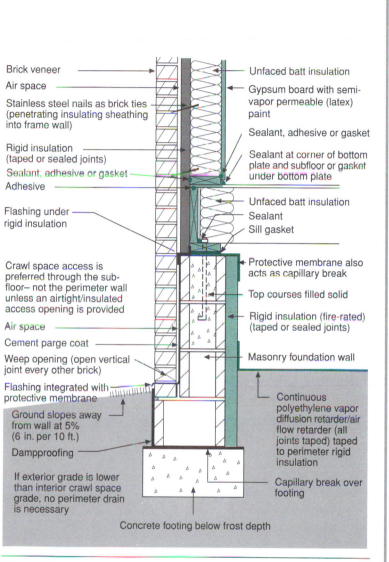

Brick veneer

Air space

Stainless steel nails as brick ties (penetrating insulating sheathing into frame wall)

Rigid insulation (taped or sealed joints)

Sealant, adhesive or gasket

Adhesive

Flashing under rigid insulation

Crawl space access is preferred through the subfloor– not the perimeter wall unless an airtight/insulated access opening is provided

Air space

Cement parge coat

Weep opening (open vertical joint every other brick)

Flashing integrated with protective membrane

Ground slopes away from wall at 5% (6 in. per 10 ft.)

Dampproofing

If exterior grade is lower than interior crawl space grade, no perimeter drain is necessary

Concrete footing below frost depth

Unfaced batt insulation

Gypsum board with semi-vapor permeable (latex) paint

Sealant, adhesive or gasket

Sealant at corner of bottom plate and subfloor or gasket under bottom plate

Unfaced batt insulation

Sealant

Sill gasket

Protective membrane also acts as capillary break

Top courses filled solid

Rigid insulation (fire-rated) (taped or sealed joints)

Masonry foundation wall

Continuous polyethylene vapor diffusion retarder/air flow retarder (all joints taped) taped to perimeter rigid insulation

Capillary break over footing

4

Foundations

Figure 4.16
Internally Insulated Masonry Crawl Space — Brick Veneer

- Masonry wall cold, can dry to the exterior; low likelihood of mold
- Protective membrane acts as termite barrier
- Rigid insulation must be fire-rated if it is left exposed on the interior
- Capillary break over footing can be omitted if sufficient drying to exterior is provided above grade
- Airspace behind brick veneer can be as small as ³/₈", 1" is typical

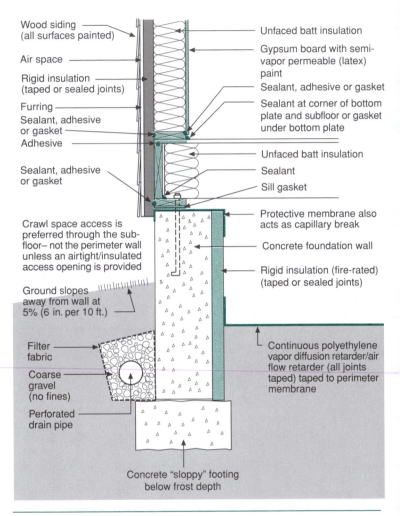

Wood siding (all surfaces painted)

Air space

Rigid insulation (taped or sealed joints)

Furring

Sealant, adhesive or gasket

Adhesive

Sealant, adhesive or gasket

Unfaced batt insulation

Gypsum board with semi-vapor permeable (latex) paint

Sealant, adhesive or gasket

Sealant at corner of bottom plate and subfloor or gasket under bottom plate

Unfaced batt insulation

Sealant

Sill gasket

Crawl space access is preferred through the sub-floor– not the perimeter wall unless an airtight/insulated access opening is provided

Ground slopes away from wall at 5% (6 in. per 10 ft.)

Filter fabric

Coarse gravel (no fines)

Perforated drain pipe

Protective membrane also acts as capillary break

Concrete foundation wall

Rigid insulation (fire-rated) (taped or sealed joints)

Continuous polyethylene vapor diffusion retarder/air flow retarder (all joints taped) taped to perimeter membrane

Concrete "sloppy" footing below frost depth

Figure 4.17
Internally Insulated Concrete Crawl Space — Wood Siding
- Concrete wall cold, can dry to exterior; low likelihood of mold
- Protective membrane acts as termite barrier
- Perimeter drain is necessary since interior grade is lower than exterior grade
- Rigid insulation must be fire-rated if it is left exposed on the interior
- Capillary break over footing can be omitted if sufficient drying to exterior is provided

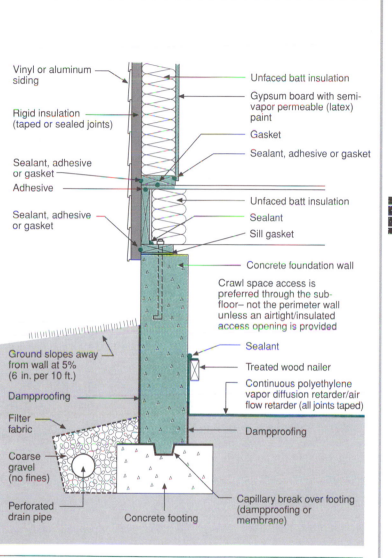

4

Foundations

Vinyl or aluminum siding

Rigid insulation (taped or sealed joints)

Sealant, adhesive or gasket

Adhesive

Sealant, adhesive or gasket

Unfaced batt insulation

Gypsum board with semi-vapor permeable (latex) paint

Gasket

Sealant, adhesive or gasket

Unfaced batt insulation

Sealant

Sill gasket

Concrete foundation wall

Crawl space access is preferred through the sub-floor– not the perimeter wall unless an airtight/insulated access opening is provided

Ground slopes away from wall at 5% (6 in. per 10 ft.)

Dampproofing

Filter fabric

Coarse gravel (no fines)

Perforated drain pipe

Concrete footing

Sealant

Treated wood nailer

Continuous polyethylene vapor diffusion retarder/air flow retarder (all joints taped)

Dampproofing

Capillary break over footing (dampproofing or membrane)

Figure 4.18
Externally Insulated Concrete Crawl Space — Vinyl or Aluminum Siding
- Concrete wall can dry to the interior and exterior; extremely low likelihood of mold
- Perimeter drain is necessary since interior grade is lower than exterior grade

4

Foundations

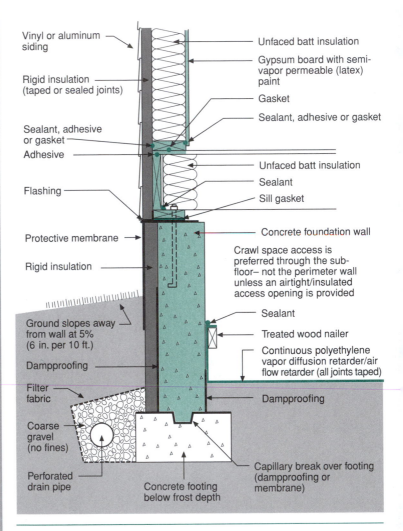

Vinyl or aluminum siding

Rigid insulation (taped or sealed joints)

Sealant, adhesive or gasket

Adhesive

Flashing

Protective membrane

Rigid insulation

Ground slopes away from wall at 5% (6 in. per 10 ft.)

Dampproofing

Filter fabric

Coarse gravel (no fines)

Perforated drain pipe

Concrete footing below frost depth

Unfaced batt insulation

Gypsum board with semi-vapor permeable (latex) paint

Gasket

Sealant, adhesive or gasket

Unfaced batt insulation

Sealant

Sill gasket

Concrete foundation wall

Crawl space access is preferred through the sub-floor— not the perimeter wall unless an airtight/insulated access opening is provided

Sealant

Treated wood nailer

Continuous polyethylene vapor diffusion retarder/air flow retarder (all joints taped)

Dampproofing

Capillary break over footing (dampproofing or membrane)

Figure 4.19
Externally Insulated Concrete Crawl Space — Vinyl or Aluminum Siding

- Concrete wall warm, can dry to the interior; extremely low likelihood of mold
- Protective membrane acts as termite barrier
- Perimeter drain is necessary since interior grade is lower than exterior grade
- Protective membrane can be adhesive-backed roll roofing or other UV resistant materials. Below grade sheet waterproofing or ice-dam protection membranes can also be used if protected from UV exposure by using aluminum sheet stock or other alternative materials.

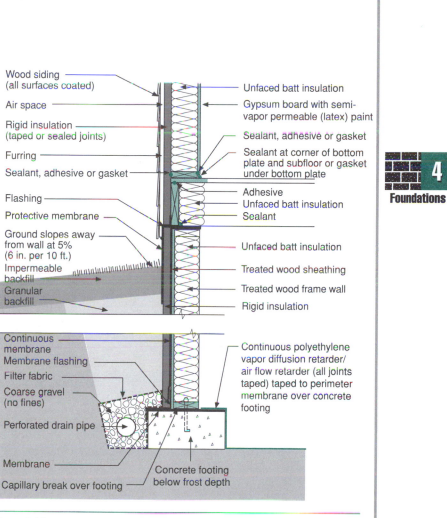

4

Foundations

Figure 4.20
Externally Insulated Wood Crawl Space — Wood Siding

- Wood frame warm, can dry to interior; low likelihood of mold
- Protective membrane acts as termite barrier
- Protective membrane can be adhesive-backed roll roofing or other UV resistant materials. Below grade sheet waterproofing or ice-dam protection membranes can also be used if protected from UV exposure by using aluminum sheet stock or other alternative materials.

4

Foundations

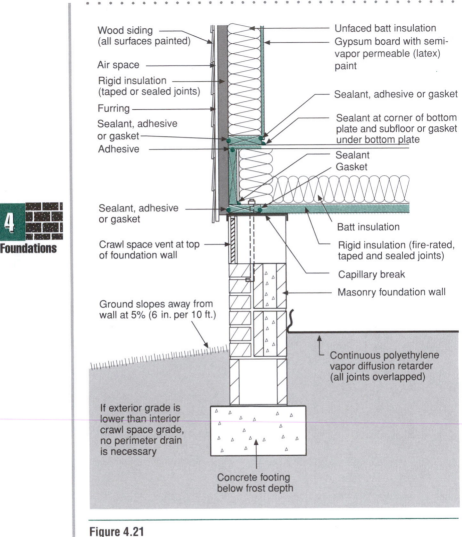

Wood siding (all surfaces painted)

Air space

Rigid insulation (taped or sealed joints)

Furring

Sealant, adhesive or gasket

Adhesive

Unfaced batt insulation

Gypsum board with semi-vapor permeable (latex) paint

Sealant, adhesive or gasket

Sealant at corner of bottom plate and subfloor or gasket under bottom plate

Sealant

Gasket

Sealant, adhesive or gasket

Crawl space vent at top of foundation wall

Batt insulation

Rigid insulation (fire-rated, taped and sealed joints)

Capillary break

Masonry foundation wall

Ground slopes away from wall at 5% (6 in. per 10 ft.)

Continuous polyethylene vapor diffusion retarder (all joints overlapped)

If exterior grade is lower than interior crawl space grade, no perimeter drain is necessary

Concrete footing below frost depth

Figure 4.21
Vented Masonry Crawl Space — Wood Siding

- Rigid impermeable insulating sheathing protects underside of floor assembly from wetting during summer. Structural wood beams must also be similarly protected.
- Rigid insulation must be fire-rated if it is left exposed under the floor framing in the crawl space
- Band joist assembly must be tight or entry of outside air will compromise the effectiveness of the floor cavity insulation
- Penetrations in bottom of interior and exterior partition walls should also be sealed to provide a degree of redundancy to the primary air flow retarder (the rigid insulation and band joist assembly)

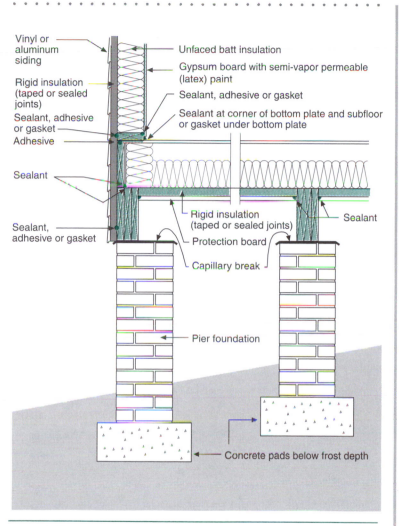

Vinyl or aluminum siding

Rigid insulation (taped or sealed joints)

Sealant, adhesive or gasket

Adhesive

Sealant

Sealant, adhesive or gasket

Unfaced batt insulation

Gypsum board with semi-vapor permeable (latex) paint

Sealant, adhesive or gasket

Sealant at corner of bottom plate and subfloor or gasket under bottom plate

Rigid insulation (taped or sealed joints)

Sealant

Protection board

Capillary break

Pier foundation

Concrete pads below frost depth

4

Foundations

Figure 4.22
Pier Foundation — Vinyl or Aluminum Siding

- Rigid impermeable insulating sheathing protects underside of floor assembly from wetting during summer
- Floor cavity insulation held down in contact with rigid insulation – air space above insulation provides warm floor during winter
- Band joist assembly must be tight or entry of outside air will compromise the effectiveness of the floor cavity insulation
- Penetrations in bottom of interior and exterior partition walls should also be sealed to provide a degree of redundancy to the primary air flow retarder (the rigid insulation and band joist assembly)

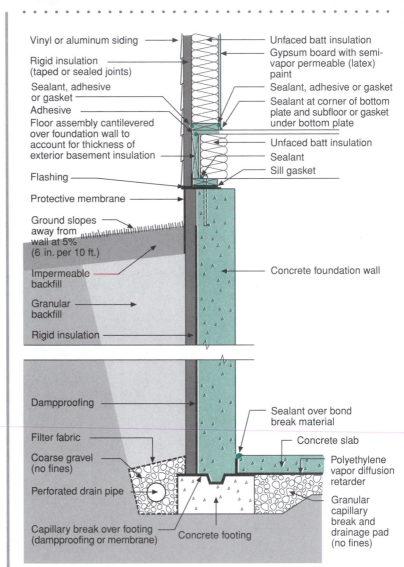

Vinyl or aluminum siding

Rigid insulation
(taped or sealed joints)

Sealant, adhesive
or gasket

Adhesive

Floor assembly cantilevered
over foundation wall to
account for thickness of
exterior basement insulation

Flashing

Protective membrane

Ground slopes
away from
wall at 5%
(6 in. per 10 ft.)

Impermeable
backfill

Granular
backfill

Rigid insulation

Damproofing

Filter fabric

Coarse gravel
(no fines)

Perforated drain pipe

Capillary break over footing
(damproofing or membrane)

Unfaced batt insulation

Gypsum board with semi-
vapor permeable (latex)
paint

Sealant, adhesive or gasket

Sealant at corner of bottom
plate and subfloor or gasket
under bottom plate

Unfaced batt insulation

Sealant

Sill gasket

Concrete foundation wall

Sealant over bond
break material

Concrete slab

Polyethylene
vapor diffusion
retarder

Granular
capillary
break and
drainage pad
(no fines)

Concrete footing

Figure 4.23
Externally Insulated Concrete Basement — Vinyl or Aluminum Siding

- Concrete wall warm, can dry to the interior; extremely low likelihood of mold
- Basement floor slab can dry to the interior
- Protective membrane acts as termite barrier
- Protective membrane can be adhesive-backed roll roofing or other UV resistant
  materials. Below grade sheet waterproofing or ice-dam protection membranes
  can also be used if protected from UV exposure by using aluminum sheet stock
  or other alternative materials.

4

Foundations

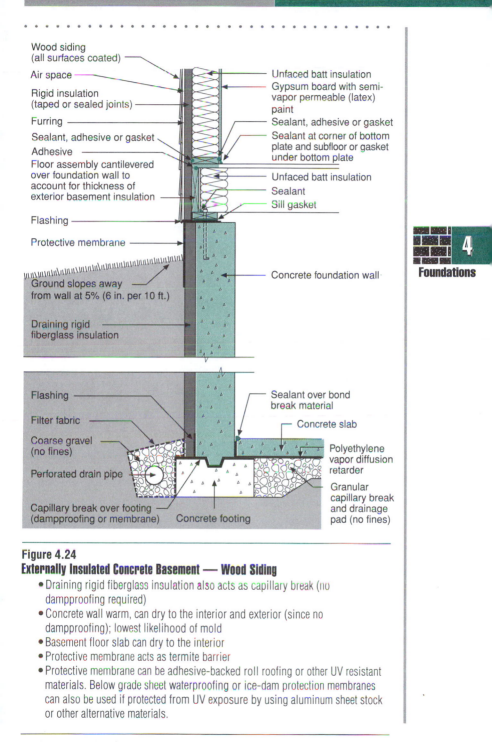

Wood siding
(all surfaces coated)

Air space

Rigid insulation
(taped or sealed joints)

Furring

Sealant, adhesive or gasket

Adhesive

Floor assembly cantilevered
over foundation wall to
account for thickness of
exterior basement insulation

Flashing

Protective membrane

Ground slopes away
from wall at 5% (6 in. per 10 ft.)

Draining rigid
fiberglass insulation

Unfaced batt insulation

Gypsum board with semi-
vapor permeable (latex)
paint

Sealant, adhesive or gasket

Sealant at corner of bottom
plate and subfloor or gasket
under bottom plate

Unfaced batt insulation

Sealant

Sill gasket

Concrete foundation wall

Flashing

Filter fabric

Coarse gravel
(no fines)

Perforated drain pipe

Capillary break over footing
(dampproofing or membrane)

Sealant over bond
break material

Concrete slab

Polyethylene
vapor diffusion
retarder

Granular
capillary break
and drainage
pad (no fines)

Concrete footing

4

Foundations

Figure 4.24
Externally Insulated Concrete Basement — Wood Siding

- Draining rigid fiberglass insulation also acts as capillary break (no
  dampproofing required)
- Concrete wall warm, can dry to the interior and exterior (since no
  dampproofing); lowest likelihood of mold
- Basement floor slab can dry to the interior
- Protective membrane acts as termite barrier
- Protective membrane can be adhesive-backed roll roofing or other UV resistant
  materials. Below grade sheet waterproofing or ice-dam protection membranes
  can also be used if protected from UV exposure by using aluminum sheet stock
  or other alternative materials.

4

Foundations

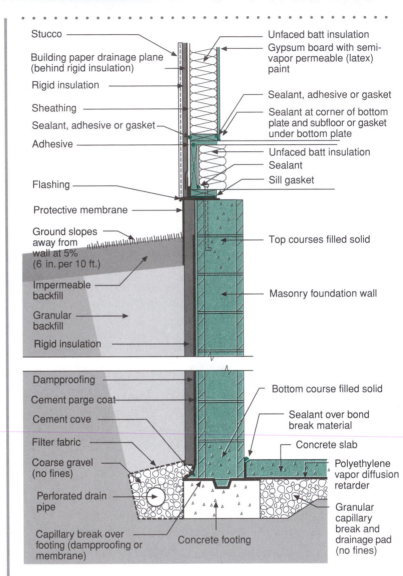

Stucco

Building paper drainage plane (behind rigid insulation)

Rigid insulation

Sheathing

Sealant, adhesive or gasket

Adhesive

Flashing

Protective membrane

Ground slopes away from wall at 5% (6 in. per 10 ft.)

Impermeable backfill

Granular backfill

Rigid insulation

Dampproofing

Cement parge coat

Cement cove

Filter fabric

Coarse gravel (no fines)

Perforated drain pipe

Capillary break over footing (dampproofing or membrane)

Unfaced batt insulation

Gypsum board with semi-vapor permeable (latex) paint

Sealant, adhesive or gasket

Sealant at corner of bottom plate and subfloor or gasket under bottom plate

Unfaced batt insulation

Sealant

Sill gasket

Top courses filled solid

Masonry foundation wall

Bottom course filled solid

Sealant over bond break material

Concrete slab

Polyethylene vapor diffusion retarder

Granular capillary break and drainage pad (no fines)

Concrete footing

Concrete slab

Figure 4.25
Externally Insulated Masonry Basement — Stucco

- Masonry wall warm, can dry to the interior; extremely low likelihood of mold
- Basement floor slab can dry to the interior
- Protective membrane acts as termite barrier
- Protective membrane can be adhesive-backed roll roofing or other UV resistant materials. Below grade sheet waterproofing or ice-dam protection membranes can also be used if protected from UV exposure by using aluminum sheet stock or other alternative materials.

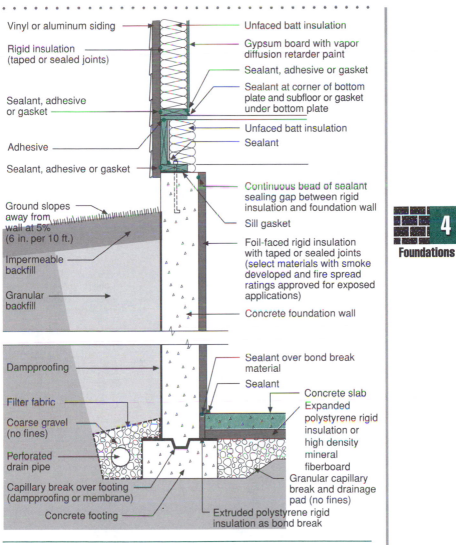

Vinyl or aluminum siding

Rigid insulation (taped or sealed joints)

Sealant, adhesive or gasket

Adhesive

Sealant, adhesive or gasket

Unfaced batt insulation

Gypsum board with vapor diffusion retarder paint

Sealant, adhesive or gasket

Sealant at corner of bottom plate and subfloor or gasket under bottom plate

Unfaced batt insulation

Sealant

Continuous bead of sealant sealing gap between rigid insulation and foundation wall

Sill gasket

Ground slopes away from wall at 5% (6 in. per 10 ft.)

Impermeable backfill

Granular backfill

Foil-faced rigid insulation with taped or sealed joints (select materials with smoke developed and fire spread ratings approved for exposed applications)

Concrete foundation wall

Dampproofing

Filter fabric

Coarse gravel (no fines)

Perforated drain pipe

Capillary break over footing (dampproofing or membrane)

Concrete footing

Sealant over bond break material

Sealant

Concrete slab

Expanded polystyrene rigid insulation or high density mineral fiberboard

Granular capillary break and drainage pad (no fines)

Extruded polystyrene rigid insulation as bond break

4
Foundations

Figure 4.26
Internally Insulated Concrete Basement — Vinyl or Aluminum Siding

- Cold concrete wall must be protected from interior moisture-laden air in winter and in summer
- Basement floor slab is warm, can dry to the ground (since no sub-slab vapor diffusion retarder) as well as to the interior; lowest likelihood of mold
- Concrete wall cold, cannot dry to interior; drying only possible to exterior at above grade portion of wall
- Due to low drying potential, interior impermeable rigid insulation should not be installed until moisture from concrete foundation wall has substantially dried/ equilibrated (typically 6 months or more) otherwise mold is possible

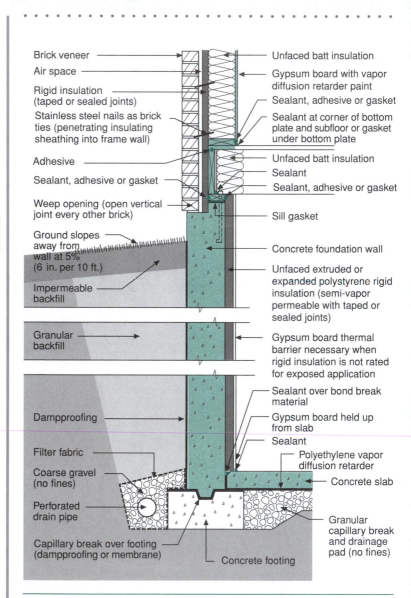

Brick veneer

Air space

Rigid insulation
(taped or sealed joints)

Stainless steel nails as brick
ties (penetrating insulating
sheathing into frame wall)

Adhesive

Sealant, adhesive or gasket

Weep opening (open vertical
joint every other brick)

Ground slopes
away from
wall at 5%
(6 in. per 10 ft.)

Impermeable
backfill

Granular
backfill

Dampproofing

Filter fabric

Coarse gravel
(no fines)

Perforated
drain pipe

Capillary break over footing
(dampproofing or membrane)

Unfaced batt insulation

Gypsum board with vapor
diffusion retarder paint

Sealant, adhesive or gasket

Sealant at corner of bottom
plate and subfloor or gasket
under bottom plate

Unfaced batt insulation

Sealant

Sealant, adhesive or gasket

Sill gasket

Concrete foundation wall

Unfaced extruded or
expanded polystyrene rigid
insulation (semi-vapor
permeable with taped or
sealed joints)

Gypsum board thermal
barrier necessary when
rigid insulation is not rated
for exposed application

Sealant over bond break
material

Gypsum board held up
from slab

Sealant

Polyethylene vapor
diffusion retarder

Concrete slab

Granular
capillary break
and drainage
pad (no fines)

Concrete footing

4

Foundations

Figure 4.27
Internally Insulated Concrete Basement — Brick Veneer
- Concrete wall cold, can only dry to the interior if interior assemblies are semi-vapor permeable; mold possible if interior assemblies do not permit drying
- Cold concrete wall must be protected from interior moisture-laden air in winter and in summer
- Basement floor slab can dry to the interior
- Airspace behind brick veneer can be as small as $3/8$", 1" is typical

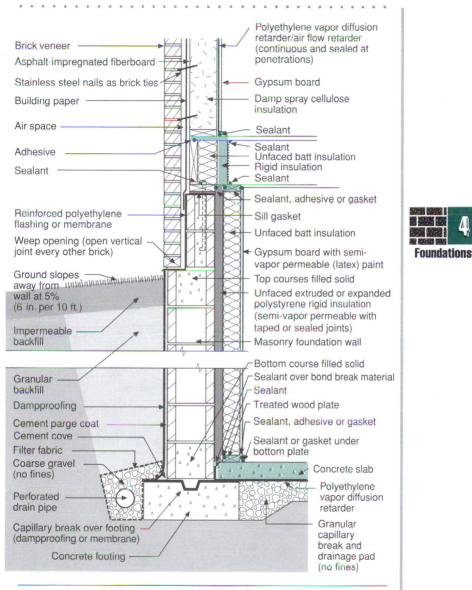

Brick veneer
Asphalt-impregnated fiberboard
Stainless steel nails as brick ties
Building paper
Air space
Adhesive
Sealant

Reinforced polyethylene
flashing or membrane
Weep opening (open vertical
joint every other brick)
Ground slopes
away from
wall at 5%
(6 in. per 10 ft.)
Impermeable
backfill

Granular
backfill
Dampproofing
Cement parge coat
Cement cove
Filter fabric
Coarse gravel
(no fines)
Perforated
drain pipe
Capillary break over footing
(dampproofing or membrane)
Concrete footing

Polyethylene vapor diffusion
retarder/air flow retarder
(continuous and sealed at
penetrations)
Gypsum board
Damp spray cellulose
insulation
Sealant
Sealant
Unfaced batt insulation
Rigid insulation
Sealant
Sealant, adhesive or gasket
Sill gasket
Unfaced batt insulation
Gypsum board with semi-
vapor permeable (latex) paint
Top courses filled solid
Unfaced extruded or expanded
polystyrene rigid insulation
(semi-vapor permeable with
taped or sealed joints)
Masonry foundation wall
Bottom course filled solid
Sealant over bond break material
Sealant
Treated wood plate
Sealant, adhesive or gasket
Sealant or gasket under
bottom plate
Concrete slab
Polyethylene
vapor diffusion
retarder
Granular
capillary
break and
drainage pad
(no fines)

4

Foundations

Figure 4.28
Internally Insulated Masonry Basement — Brick Veneer

- Masonry wall cold, can only dry to the interior if interior assemblies are vapor
  permeable; mold possible if interior assemblies do not permit drying
- Cold masonry wall must be protected from interior moisture-laden air in winter
  and in summer
- Basement floor slab can dry to the interior
- Airspace behind brick veneer can be as small as $^3/_8$", 1" is typical

4

Foundations

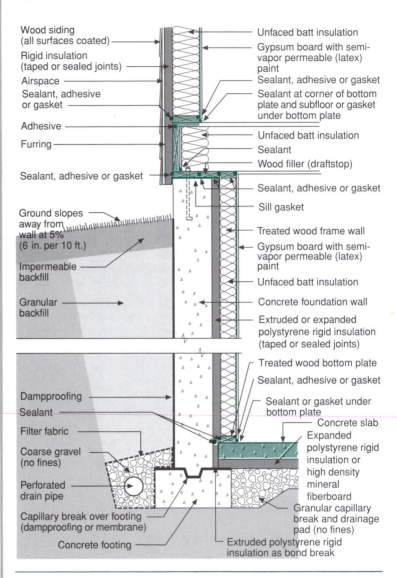

Wood siding (all surfaces coated)
Rigid insulation (taped or sealed joints)
Airspace
Sealant, adhesive or gasket
Adhesive
Furring
Sealant, adhesive or gasket
Ground slopes away from wall at 5% (6 in. per 10 ft.)
Impermeable backfill
Granular backfill

Unfaced batt insulation
Gypsum board with semi-vapor permeable (latex) paint
Sealant, adhesive or gasket
Sealant at corner of bottom plate and subfloor or gasket under bottom plate
Unfaced batt insulation
Sealant
Wood filler (draftstop)
Sealant, adhesive or gasket
Sill gasket
Treated wood frame wall
Gypsum board with semi-vapor permeable (latex) paint
Unfaced batt insulation
Concrete foundation wall
Extruded or expanded polystyrene rigid insulation (taped or sealed joints)
Treated wood bottom plate
Sealant, adhesive or gasket
Sealant or gasket under bottom plate
Concrete slab

Dampproofing
Sealant
Filter fabric
Coarse gravel (no fines)
Perforated drain pipe
Capillary break over footing (dampproofing or membrane)
Concrete footing

Expanded polystyrene rigid insulation or high density mineral fiberboard
Granular capillary break and drainage pad (no fines)
Extruded polystyrene rigid insulation as bond break

Figure 4.29
Internally Insulated Concrete Basement — Wood Siding

- Concrete wall cold, can only dry to the interior if interior assemblies are semi-vapor permeable; low likelihood of mold
- Cold concrete wall must be protected from interior moisture-laden air in winter and in summer
- Basement floor slab is warm, can dry to the ground (since there is no under slab vapor diffusion retarder) as well as to the interior; lowest likelihood of mold

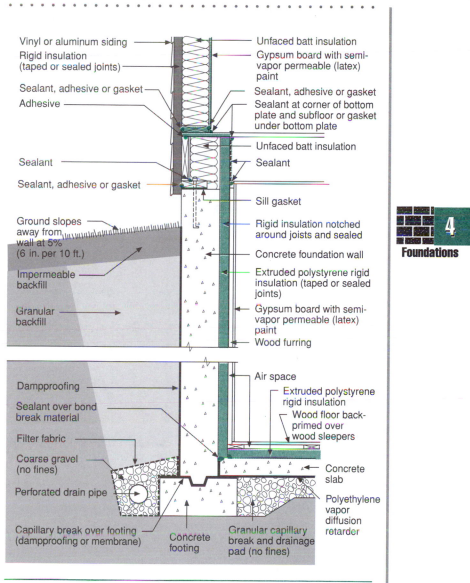

Vinyl or aluminum siding

Rigid insulation
(taped or sealed joints)

Sealant, adhesive or gasket

Adhesive

Sealant

Sealant, adhesive or gasket

Ground slopes
away from
wall at 5%
(6 in. per 10 ft.)

Impermeable
backfill

Granular
backfill

Dampproofing

Sealant over bond
break material

Filter fabric

Coarse gravel
(no fines)

Perforated drain pipe

Capillary break over footing
(dampproofing or membrane)

Unfaced batt insulation

Gypsum board with semi-
vapor permeable (latex)
paint

Sealant, adhesive or gasket

Sealant at corner of bottom
plate and subfloor or gasket
under bottom plate

Unfaced batt insulation

Sealant

Sill gasket

Rigid insulation notched
around joists and sealed

Concrete foundation wall

Extruded polystyrene rigid
insulation (taped or sealed
joints)

Gypsum board with semi-
vapor permeable (latex)
paint

Wood furring

Air space

Extruded polystyrene
rigid insulation

Wood floor back-
primed over
wood sleepers

Concrete
slab

Polyethylene
vapor
diffusion
retarder

Concrete
footing

Granular capillary
break and drainage
pad (no fines)

Foundations

4

Figure 4.30

Internally Insulated Concrete Basement — Vinyl or Aluminum Siding

- Concrete wall cold, can only dry to the interior if interior assemblies are semi-vapor permeable; low likelihood of mold
- Cold concrete wall must be protected from interior moisture-laden air in winter and in summer
- Basement floor slab can dry to the interior if interior assemblies are semi-vapor permeable; low likelihood of mold

4

Foundations

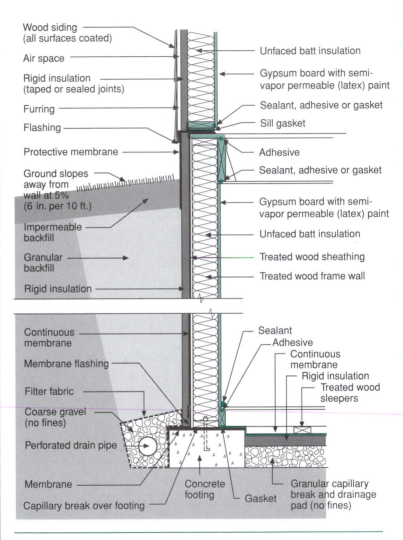

Wood siding (all surfaces coated)

Air space

Rigid insulation (taped or sealed joints)

Furring

Flashing

Protective membrane

Ground slopes away from wall at 5% (6 in. per 10 ft.)

Impermeable backfill

Granular backfill

Rigid insulation

Unfaced batt insulation

Gypsum board with semi-vapor permeable (latex) paint

Sealant, adhesive or gasket

Sill gasket

Adhesive

Sealant, adhesive or gasket

Gypsum board with semi-vapor permeable (latex) paint

Unfaced batt insulation

Treated wood sheathing

Treated wood frame wall

Continuous membrane

Membrane flashing

Filter fabric

Coarse gravel (no fines)

Perforated drain pipe

Membrane

Capillary break over footing

Concrete footing

Gasket

Sealant

Adhesive

Continuous membrane

Rigid insulation

Treated wood sleepers

Granular capillary break and drainage pad (no fines)

Figure 4.31
Externally Insulated Wood Basement — Wood Siding

- Wood frame warm, can only dry to the interior if interior finishes are vapor permeable; low likelihood of mold
- Wood floor assembly is warm, can dry to the interior; low likelihood of mold
- Protective membrane acts as termite barrier
- Protective membrane can be adhesive-backed roll roofing or other UV resistant materials. Below grade sheet waterproofing or ice-dam protection membranes can also be used if protected from UV exposure by using aluminum sheet stock or other alternative materials.

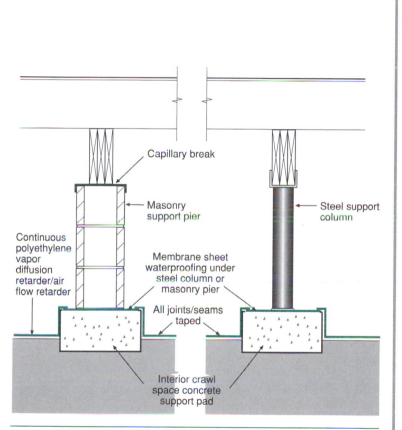

Capillary break

Masonry
support pier

Steel support
column

Continuous
polyethylene
vapor
diffusion
retarder/air
flow retarder

Membrane sheet
waterproofing under
steel column or
masonry pier

All joints/seams
taped

Interior crawl
space concrete
support pad

Figure 4.32
Air Flow Retarder Continuity at Piers

- All joints and seams in the polyethylene are taped
- Polyethylene ground cover taped to membrane sheet waterproofing at columns
  and piers

4

Foundations

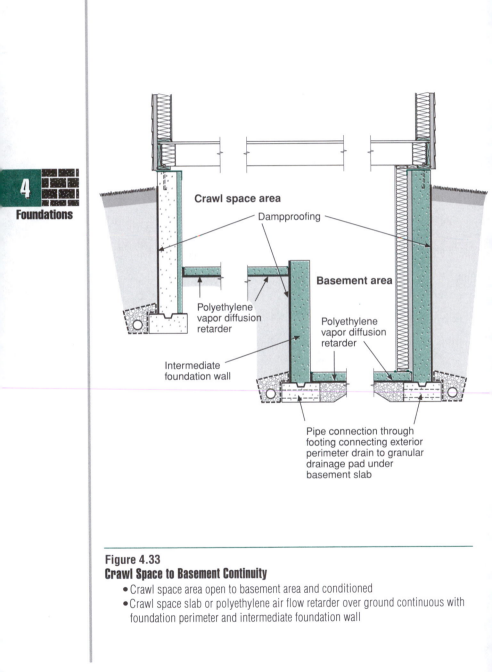

Figure 4.33
Crawl Space to Basement Continuity
- Crawl space area open to basement area and conditioned
- Crawl space slab or polyethylene air flow retarder over ground continuous with foundation perimeter and intermediate foundation wall

Framing

Three framing approaches are common in residential construction: platform frame, balloon frame, and post and beam. They can use combinations of wood, wood products, steel, masonry and concrete. They can be insulated on the inside, on the outside or in between. However, all have to:

- hold the building up
- keep the rainwater out
- keep the wind out
- keep the water vapor out
- let the water and water vapor out if it gets inside
- keep the heat in during the winter
- keep the heat out during the summer

Concerns

If someone today invented wood, it would never be approved as a building material. It burns, it rots, it has different strength properties depending on its orientation, no two pieces are alike, and most cruelly of all, it expands and contracts based on the relative humidity around it. However, despite all of these problems, wood is the material of choice when building houses. In fact, we can use wood better than we can use steel, masonry and concrete.

We can compensate more easily for wood's poor qualities than those of steel, masonry and concrete. Steel is worse in a fire than wood because it twists and bends; steel rusts more easily than wood rots; and it is expensive and difficult to compensate for its thermal inefficiencies in an exterior wall. Masonry is too expensive in most applications to use structurally, as is concrete. In an earthquake, the most dangerous buildings are made of masonry and insufficiently reinforced concrete. Needless to say, it is expensive to reinforce masonry and concrete.

We are a wood dominated industry and will likely remain so. We have learned how to work with wood over the past several hundred years to overcome its inherent deficiencies. However, despite our vast experience, we must use wood better.

We use wood inefficiently. We put too much of it in our buildings in the wrong places and in the wrong ways. And then, in the presence of all this waste, we don't put it where it is needed.

Not enough wood in the right place is obviously a problem. Oops, it fell down. But how can too much wood hurt? Well, wherever you put wood, you can't put insulation. Where you don't have insulation, you have cold spots or hot spots. Cold spots and hot spots always cause trouble. Wood is also expensive; use too much of it and you hurt your wallet.

Frame Movement

5

Framing

Remember, wood always moves. How to attach things to something that is always moving becomes real important. Gypsum board doesn't crack all by itself. It cracks because what it is attached to moves more than the gypsum can. Use too much wood in the wrong places, use too much attachment of gypsum board to wood that is moving in the wrong places, and presto, you have cracks. Lots of them. It's better to use less wood with fewer attachments. One of life's least appreciated ironies is the more you attach gypsum board to wood, the more cracks you get. Longer nails, more nail pops. More nails, more cracks. More about this in the Drywall Chapter (See Chapter 10). In the meantime, in this chapter, we will show how to eliminate as much wood as possible, by making sure we put it only where we really need it — for structure, draftstopping and firestopping.

In the past, framers have used wood much like drunken sailors on leave spend money. They felt they could never run out of wood and they put it everywhere, even where it was not needed. We do still. Look around your job site. Headers can be found in non-load bearing interior and exterior walls. Double plates are everywhere because we have not taken the time to figure out how to line up roof framing with wall and floor framing. Three stud corners to support gypsum board that doesn't need or want support. Cripples under window framing even though we hang windows. Studs on 16 in. centers rather than 24 in. centers to support gypsum board and siding that don't need it. Figures 5.1 through 5.5 and Figure 5.10 describe framing techniques that reduce wood waste, increase structural efficiency, promote thermally efficient walls and help reduce drywall cracking.

But where we really need wood, for draftstopping and firestopping we can't find it.

Rain

Another of life's ironies is that the strategy selected to keep rain out of a building will impact how it is framed. So, maybe we should decide what the strategy is before we frame? Framers are responsible for installing building paper, sheathing and windows. The rain control strategy will decide whether building papers will be used or not, whether sheathing will be taped or glued or both, and whether window openings will be wrapped or not. Figures 5.56 through 5.60 show important flashing details when using taped insulating sheathing as a drainage plane. Keeping rain out of buildings is discussed in Appendix II.

Air Flow Retarder

The strategy selected to keep outside and inside air out of the building envelope will also impact the framing approach. Framers are responsible for installing exterior housewraps, insulating sheathings, as well as the draftstops, firestops and framing used in rigid interior air flow retarders. Figures 5.6 through 5.9 and 5.11 through 5.19 show installation techniques for installing exterior insulating sheathings as exterior air flow retarders. Figures 5.20 through 5.46 show important air sealing details (draft-stopping and firestopping) that the framer must provide. Air flow retarders are discussed in Appendix III.

Framing

Moisture

The strategy selected to keep water vapor out of the building envelope yet allow moisture vapor out of the building envelope should it get in, will impact the framing approach since the approach selected will specify the type of sheathing used. Sheathings and vapor diffusion retarders are discussed in Appendix IV. In addition, if a ventilated roof strategy is to be employed, the framer must install roofing members so that air can in fact flow from soffits to vents (Figures 5.47 through 5.52).

Paint and Trim

The manner in which wood siding and wood trim is installed determines the useful service life of paint and stain coatings as well as their useful service life. Wood siding and trim should always be coated on all six surfaces and should always be installed over spacers to promote drainage and drying (Figures 5.54 and 5.56). Paints and coatings are discussed in the Painting Chapter (See Chapter 11).

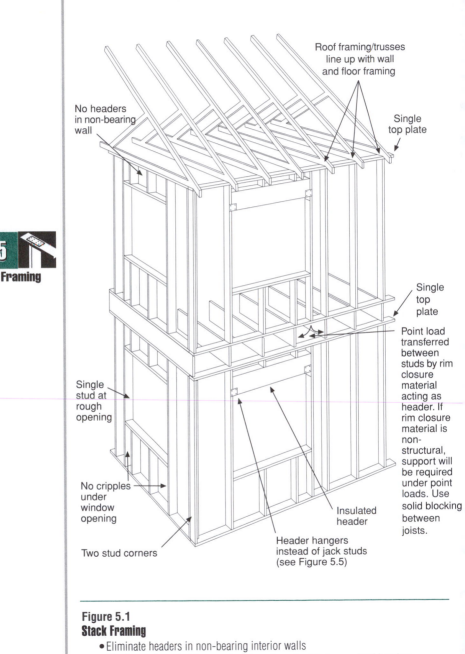

Roof framing/trusses line up with wall and floor framing

No headers in non-bearing wall

Single top plate

Single top plate

Point load transferred between studs by rim closure material acting as header. If rim closure material is non-structural, support will be required under point loads. Use solid blocking between joists.

Single stud at rough opening

No cripples under window opening

Insulated header

Two stud corners

Header hangers instead of jack studs (see Figure 5.5)

Figure 5.1
Stack Framing
- Eliminate headers in non-bearing interior walls
- Headers not needed for openings less than 4'-wide in non-load bearing exterior walls

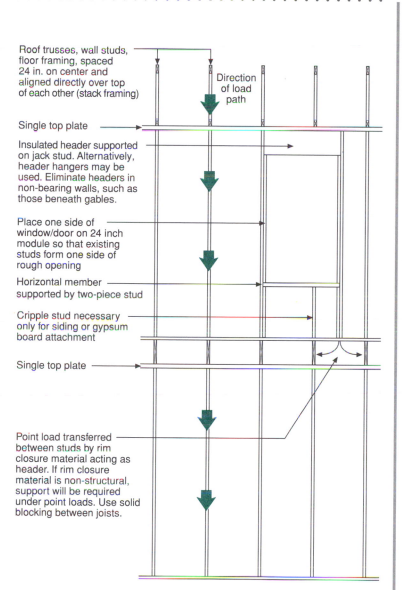

Roof trusses, wall studs, floor framing, spaced 24 in. on center and aligned directly over top of each other (stack framing)

Direction of load path

Single top plate

Insulated header supported on jack stud. Alternatively, header hangers may be used. Eliminate headers in non-bearing walls, such as those beneath gables.

Place one side of window/door on 24 inch module so that existing studs form one side of rough opening

Horizontal member supported by two-piece stud

Cripple stud necessary only for siding or gypsum board attachment

Single top plate

Point load transferred between studs by rim closure material acting as header. If rim closure material is non-structural, support will be required under point loads. Use solid blocking between joists.

5

Framing

Figure 5.2
Stack Framing Elevation View

- Where single plates are used, floor to ceiling heights are affected (97" is standard). Custom cutting dimensional studs to 94" is recommended and results in no impact on gypsum board installation.

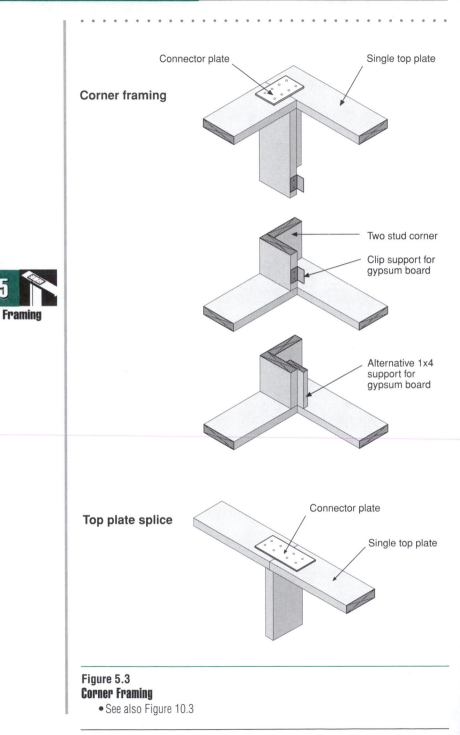

Corner framing

Connector plate

Single top plate

Two stud corner

Clip support for gypsum board

Alternative 1x4 support for gypsum board

Top plate splice

Connector plate

Single top plate

Figure 5.3
Corner Framing
• See also Figure 10.3

5
Framing

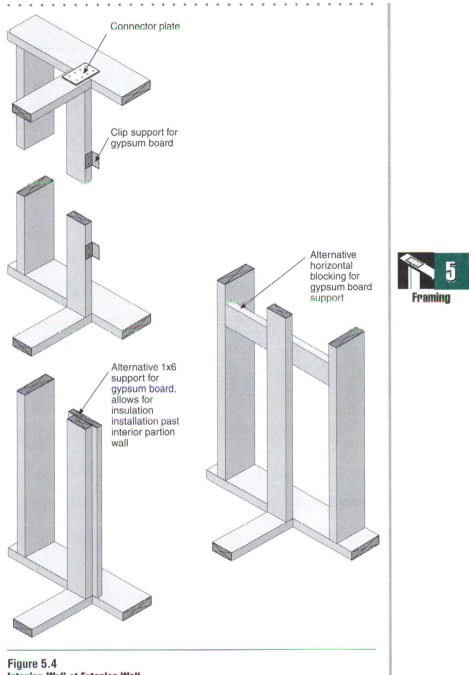

Connector plate

Clip support for gypsum board

Alternative horizontal blocking for gypsum board support

Alternative 1x6 support for gypsum board, allows for insulation installation past interior partion wall

5

Framing

Figure 5.4
Interior Wall at Exterior Wall
• See also Figures 10.3 and 10.7

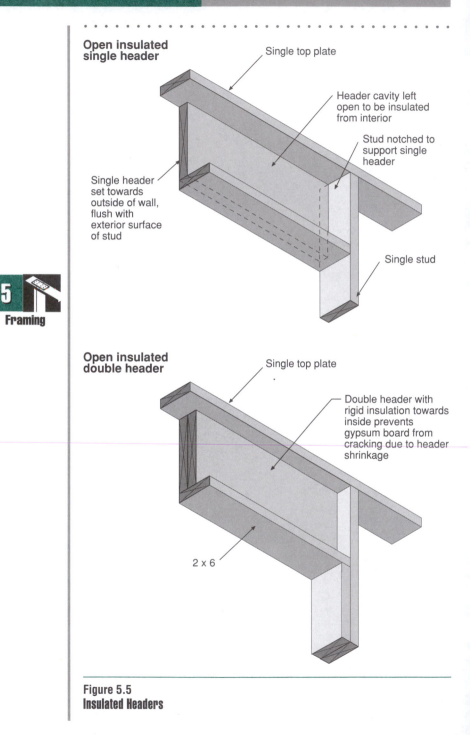

Open insulated single header

Single top plate

Header cavity left open to be insulated from interior

Stud notched to support single header

Single header set towards outside of wall, flush with exterior surface of stud

Single stud

Open insulated double header

Single top plate

Double header with rigid insulation towards inside prevents gypsum board from cracking due to header shrinkage

2 x 6

Figure 5.5
Insulated Headers

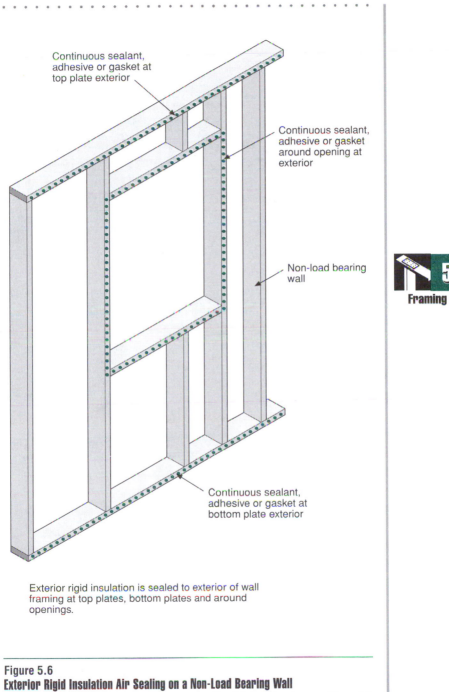

Continuous sealant, adhesive or gasket at top plate exterior

Continuous sealant, adhesive or gasket around opening at exterior

Non-load bearing wall

Continuous sealant, adhesive or gasket at bottom plate exterior

5

Framing

Exterior rigid insulation is sealed to exterior of wall framing at top plates, bottom plates and around openings.

Figure 5.6
Exterior Rigid Insulation Air Sealing on a Non-Load Bearing Wall
- A header is needed above window opening if wall is load bearing; see Figure 5.7

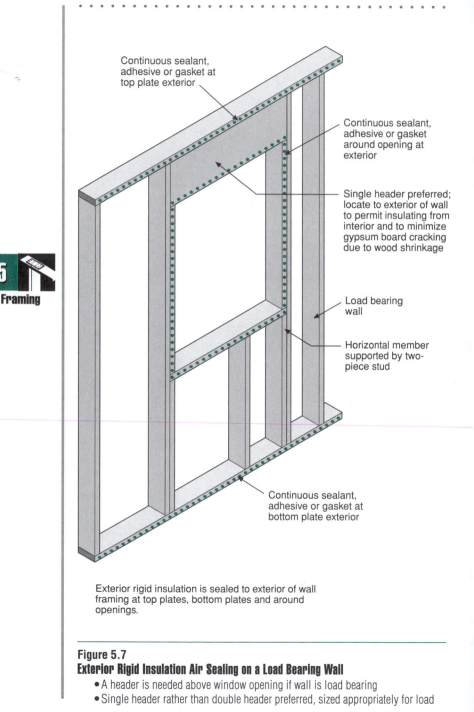

Continuous sealant, adhesive or gasket at top plate exterior

Continuous sealant, adhesive or gasket around opening at exterior

Single header preferred; locate to exterior of wall to permit insulating from interior and to minimize gypsum board cracking due to wood shrinkage

Load bearing wall

Horizontal member supported by two-piece stud

Continuous sealant, adhesive or gasket at bottom plate exterior

5

Framing

Exterior rigid insulation is sealed to exterior of wall framing at top plates, bottom plates and around openings.

Figure 5.7
Exterior Rigid Insulation Air Sealing on a Load Bearing Wall
- A header is needed above window opening if wall is load bearing
- Single header rather than double header preferred, sized appropriately for load

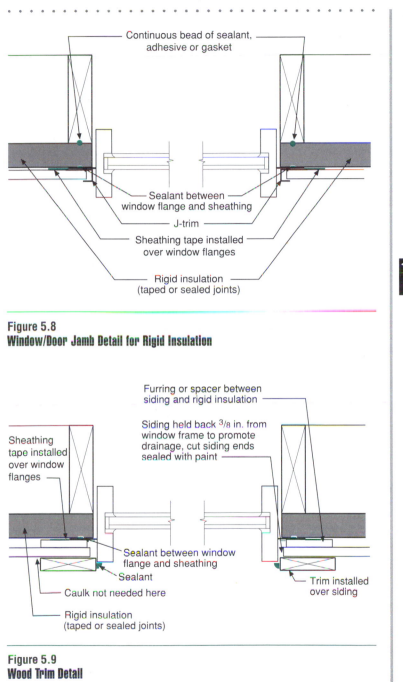

Continuous bead of sealant, adhesive or gasket

Sealant between window flange and sheathing

J-trim

Sheathing tape installed over window flanges

Rigid insulation (taped or sealed joints)

Figure 5.8
Window/Door Jamb Detail for Rigid Insulation

Furring or spacer between siding and rigid insulation

Siding held back 3/8 in. from window frame to promote drainage, cut siding ends sealed with paint

Sheathing tape installed over window flanges

Sealant between window flange and sheathing

Sealant

Caulk not needed here

Rigid insulation (taped or sealed joints)

Trim installed over siding

Figure 5.9
Wood Trim Detail

- Wood window trim nailed over siding to promote water drainage and drying

Framing

5

Each wall should have pairs of cross braces, crossing from top to bottom in opposite directions.

Cross bracing tied into top and bottom plates

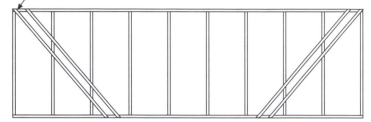

Wrapping metal braces over top of top plates and under bottom plates and fastening down into top plate or up into bottom plate significantly improves shear resistance

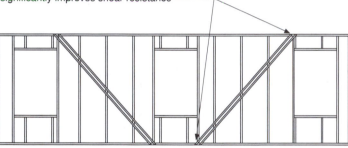

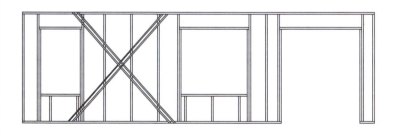

Figure 5.10
Cross Bracing

- Structural requirements and capacity should be determined on a case-by-case basis
- Interior partitions can also be designed to provide shear resistance

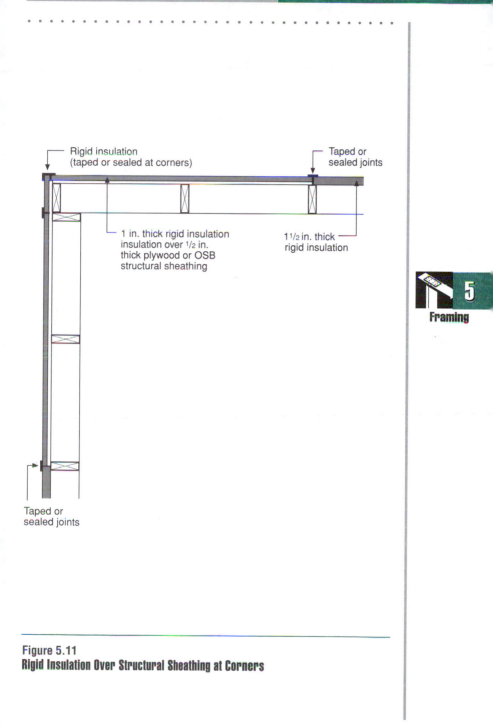

Rigid insulation
(taped or sealed at corners)

Taped or
sealed joints

1 in. thick rigid insulation
insulation over 1/2 in.
thick plywood or OSB
structural sheathing

1 1/2 in. thick
rigid insulation

5

Framing

Taped or
sealed joints

**Figure 5.11
Rigid Insulation Over Structural Sheathing at Corners**

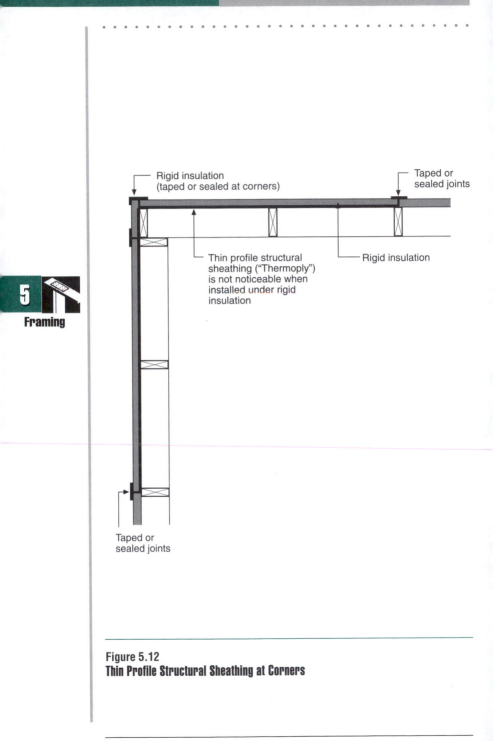

Rigid insulation
(taped or sealed at corners)

Taped or
sealed joints

Thin profile structural
sheathing ("Thermoply")
is not noticeable when
installed under rigid
insulation

Rigid insulation

Taped or
sealed joints

5

Framing

Figure 5.12
Thin Profile Structural Sheathing at Corners

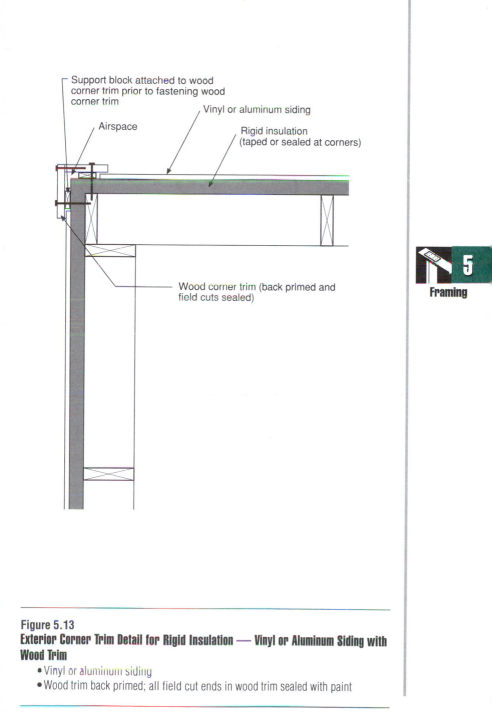

Support block attached to wood
corner trim prior to fastening wood
corner trim

Airspace

Vinyl or aluminum siding

Rigid insulation
(taped or sealed at corners)

Wood corner trim (back primed and
field cuts sealed)

5

Framing

Figure 5.13
**Exterior Corner Trim Detail for Rigid Insulation — Vinyl or Aluminum Siding with
Wood Trim**
- Vinyl or aluminum siding
- Wood trim back primed; all field cut ends in wood trim sealed with paint

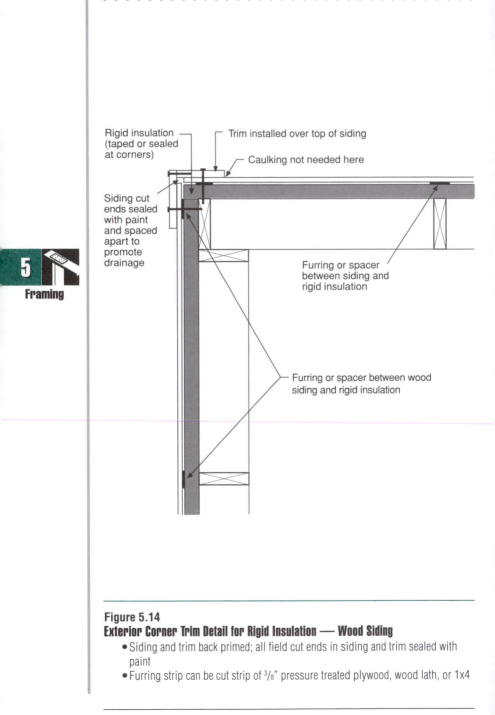

5

Framing

Rigid insulation
(taped or sealed
at corners)

Trim installed over top of siding

Caulking not needed here

Siding cut
ends sealed
with paint
and spaced
apart to
promote
drainage

Furring or spacer
between siding and
rigid insulation

Furring or spacer between wood
siding and rigid insulation

Figure 5.14
Exterior Corner Trim Detail for Rigid Insulation — Wood Siding

- Siding and trim back primed; all field cut ends in siding and trim sealed with
  paint
- Furring strip can be cut strip of $^3/_8$" pressure treated plywood, wood lath, or 1x4

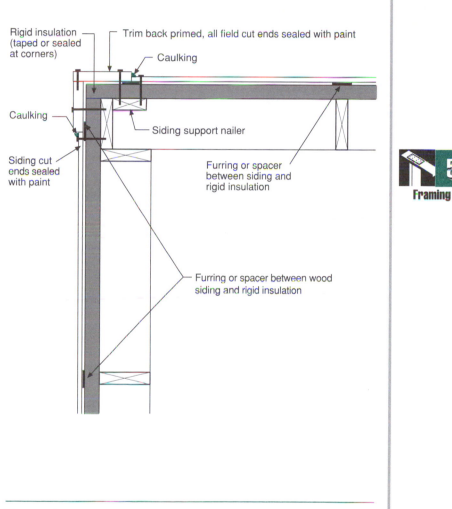

Rigid insulation
(taped or sealed
at corners)

Trim back primed, all field cut ends sealed with paint

Caulking

Caulking

Siding support nailer

Siding cut
ends sealed
with paint

Furring or spacer
between siding and
rigid insulation

Furring or spacer between wood
siding and rigid insulation

5

Framing

Figure 5.15
Exterior Corner Trim Detail for Rigid Insulation — Wood Siding

- Siding and trim back primed; all field cut ends in siding and trim sealed with paint
- Furring strip can be cut strip of $^3/_8$" pressure treated plywood, wood lath, or 1x4

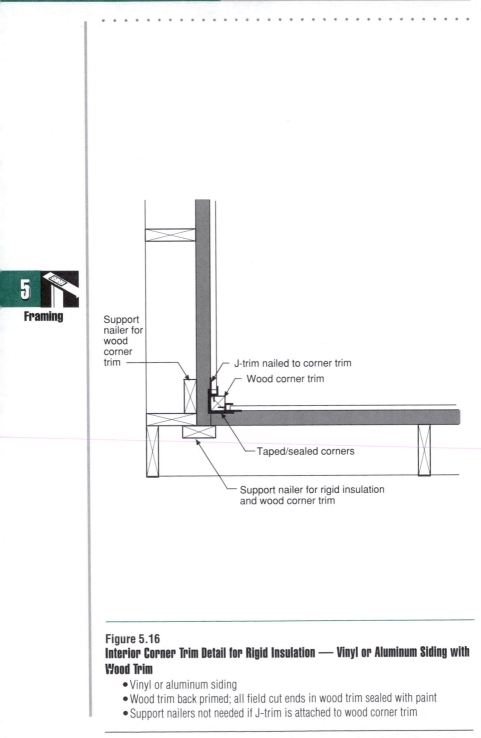

5

Framing

Support nailer for wood corner trim

J-trim nailed to corner trim

Wood corner trim

Taped/sealed corners

Support nailer for rigid insulation and wood corner trim

Figure 5.16
Interior Corner Trim Detail for Rigid Insulation — Vinyl or Aluminum Siding with Wood Trim
- Vinyl or aluminum siding
- Wood trim back primed; all field cut ends in wood trim sealed with paint
- Support nailers not needed if J-trim is attached to wood corner trim

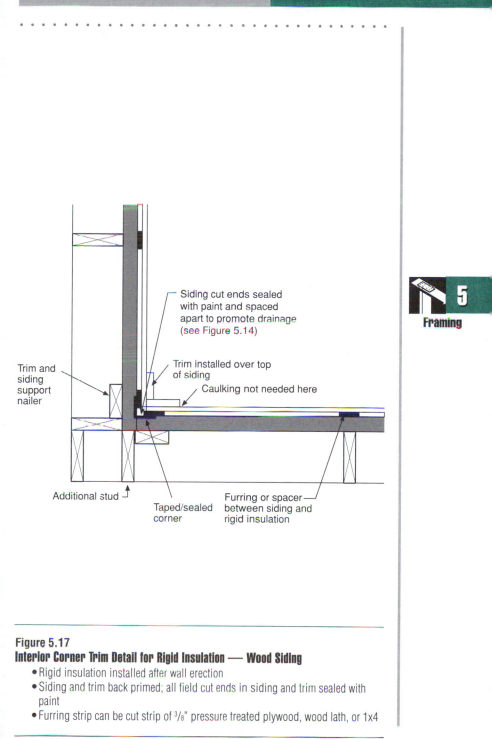

Siding cut ends sealed with paint and spaced apart to promote drainage (see Figure 5.14)

Trim installed over top of siding

Caulking not needed here

Trim and siding support nailer

Additional stud

Taped/sealed corner

Furring or spacer between siding and rigid insulation

5

Framing

Figure 5.17
Interior Corner Trim Detail for Rigid Insulation — Wood Siding

- Rigid insulation installed after wall erection
- Siding and trim back primed; all field cut ends in siding and trim sealed with paint
- Furring strip can be cut strip of $^3/_8$" pressure treated plywood, wood lath, or 1x4

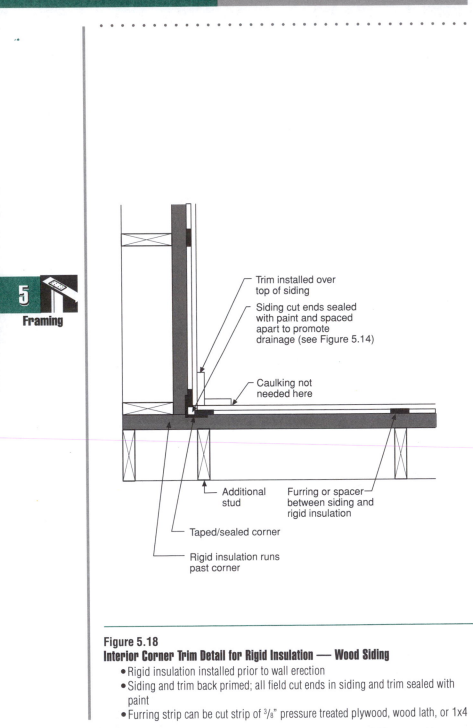

Trim installed over top of siding

Siding cut ends sealed with paint and spaced apart to promote drainage (see Figure 5.14)

Caulking not needed here

Additional stud

Furring or spacer between siding and rigid insulation

Taped/sealed corner

Rigid insulation runs past corner

Figure 5.18
Interior Corner Trim Detail for Rigid Insulation — Wood Siding
- Rigid insulation installed prior to wall erection
- Siding and trim back primed; all field cut ends in siding and trim sealed with paint
- Furring strip can be cut strip of $^3/_8$" pressure treated plywood, wood lath, or 1x4

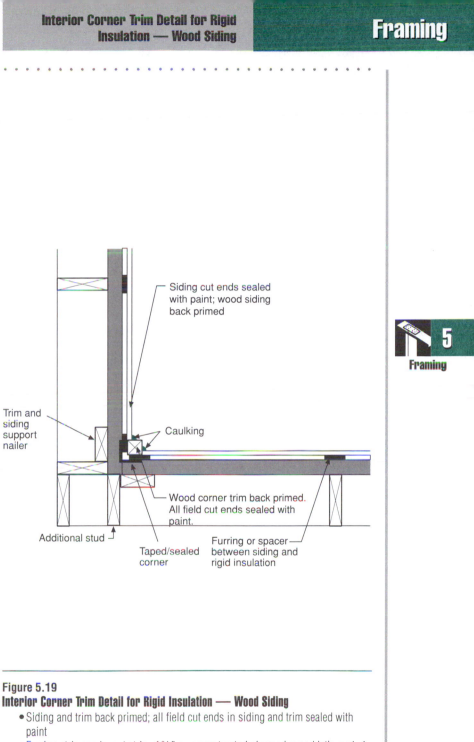

5

Framing

Figure 5.19
Interior Corner Trim Detail for Rigid Insulation — Wood Siding

- Siding and trim back primed; all field cut ends in siding and trim sealed with paint
- Furring strip can be cut strip of $^3/_8$" pressure treated plywood, wood lath, or 1x4

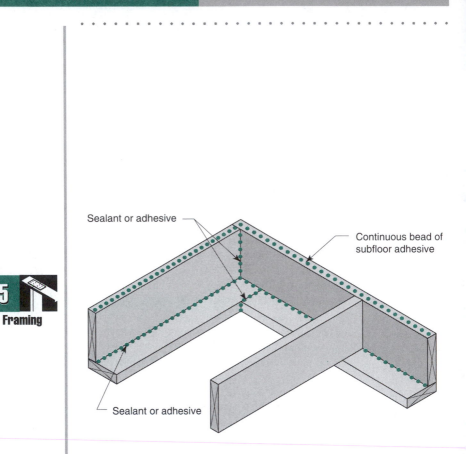

Sealant or adhesive

Continuous bead of
subfloor adhesive

Sealant or adhesive

Figure 5.20
Rim Joist/Band Joist Rim Closure
- Gaskets can be used in place of sealants or adhesives

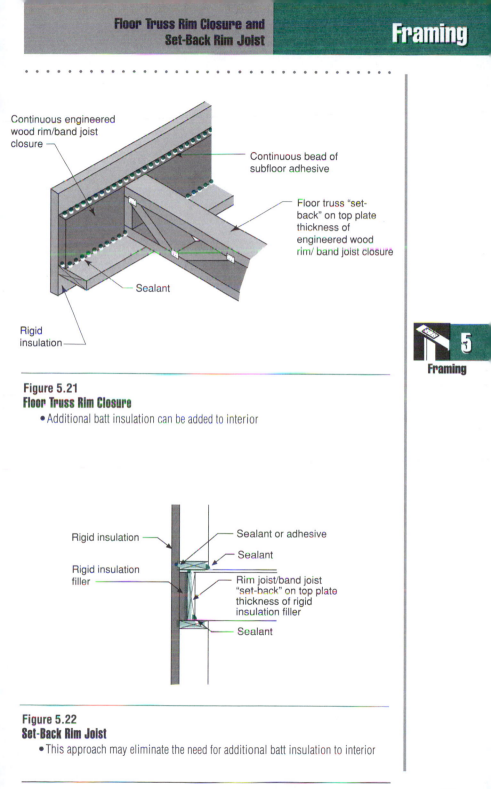

Continuous engineered wood rim/band joist closure

Continuous bead of subfloor adhesive

Floor truss "set-back" on top plate thickness of engineered wood rim/ band joist closure

Sealant

Rigid insulation

5

Framing

Figure 5.21
Floor Truss Rim Closure
- Additional batt insulation can be added to interior

Rigid insulation

Sealant or adhesive

Sealant

Rigid insulation filler

Rim joist/band joist "set-back" on top plate thickness of rigid insulation filler

Sealant

Figure 5.22
Set-Back Rim Joist
- This approach may eliminate the need for additional batt insulation to interior

5

Framing

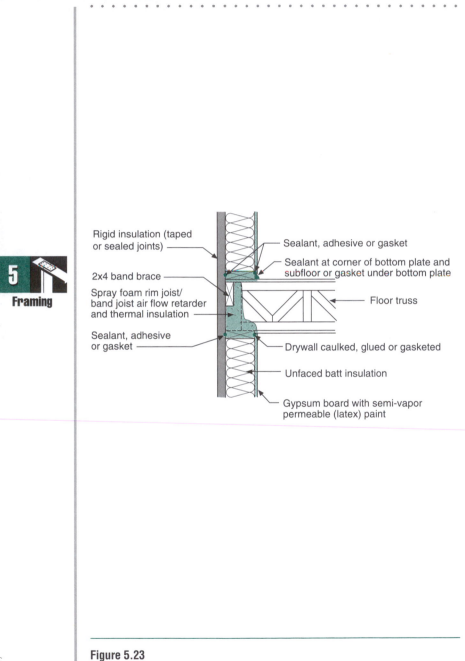

Rigid insulation (taped or sealed joints)

2x4 band brace

Spray foam rim joist/ band joist air flow retarder and thermal insulation

Sealant, adhesive or gasket

Sealant, adhesive or gasket

Sealant at corner of bottom plate and subfloor or gasket under bottom plate

Floor truss

Drywall caulked, glued or gasketed

Unfaced batt insulation

Gypsum board with semi-vapor permeable (latex) paint

Figure 5.23
Band Brace/Floor Truss Closure

- Spray foam is used to provide air flow retarder continuity across band brace region

Framing

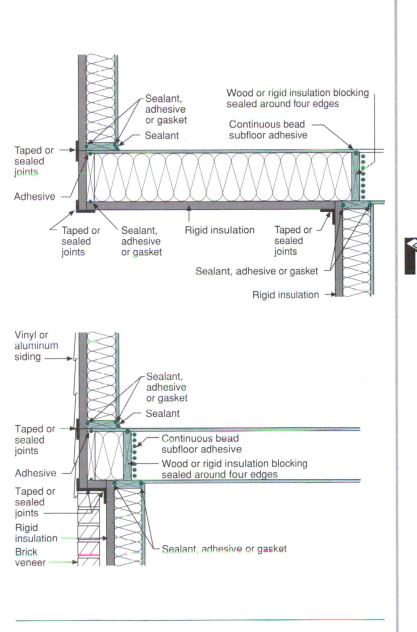

Sealant, adhesive or gasket

Sealant

Wood or rigid insulation blocking sealed around four edges

Continuous bead subfloor adhesive

Taped or sealed joints

Adhesive

Taped or sealed joints

Sealant, adhesive or gasket

Rigid insulation

Taped or sealed joints

Sealant, adhesive or gasket

Rigid insulation

Vinyl or aluminum siding

Sealant, adhesive or gasket

Sealant

Taped or sealed joints

Continuous bead subfloor adhesive

Wood or rigid insulation blocking sealed around four edges

Adhesive

Taped or sealed joints

Rigid insulation

Brick veneer

Sealant, adhesive or gasket

Figure 5.24
Cantilevered Floors

• Floor cavity insulation installed by framers prior to rigid insulation installation

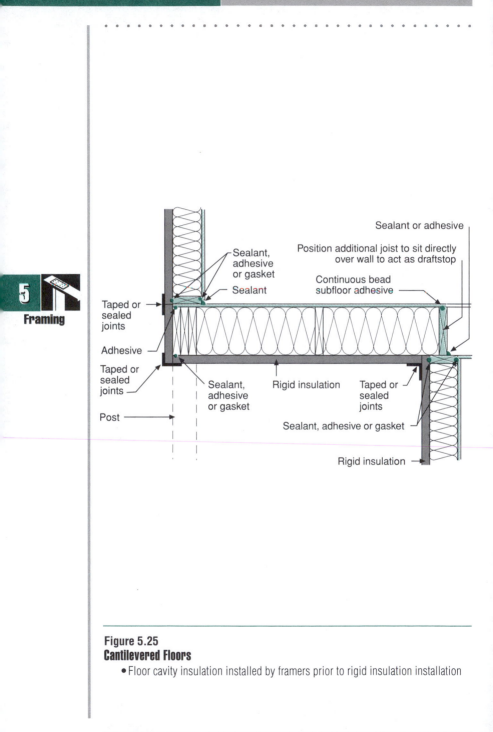

5

Framing

Sealant or adhesive

Position additional joist to sit directly over wall to act as draftstop

Continuous bead subfloor adhesive

Sealant, adhesive or gasket

Sealant

Taped or sealed joints

Adhesive

Taped or sealed joints

Post

Sealant, adhesive or gasket

Rigid insulation

Taped or sealed joints

Sealant, adhesive or gasket

Rigid insulation

Figure 5.25
Cantilevered Floors
- Floor cavity insulation installed by framers prior to rigid insulation installation

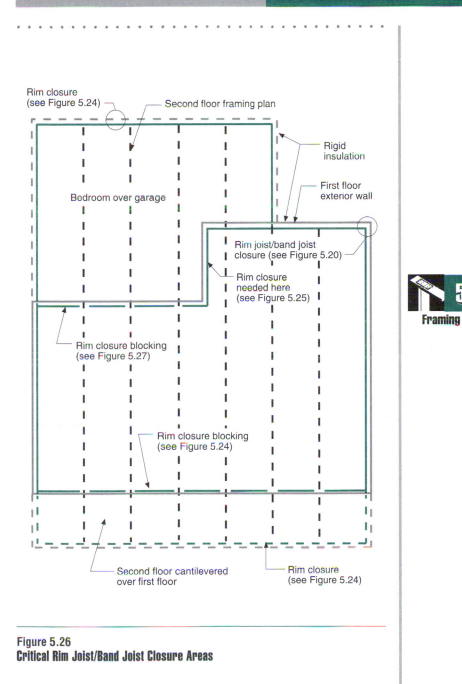

Rim closure
(see Figure 5.24)

Second floor framing plan

Rigid
insulation

First floor
exterior wall

Bedroom over garage

Rim joist/band joist
closure (see Figure 5.20)

Rim closure
needed here
(see Figure 5.25)

5

Framing

Rim closure blocking
(see Figure 5.27)

Rim closure blocking
(see Figure 5.24)

Second floor cantilevered
over first floor

Rim closure
(see Figure 5.24)

Figure 5.26
Critical Rim Joist/Band Joist Closure Areas

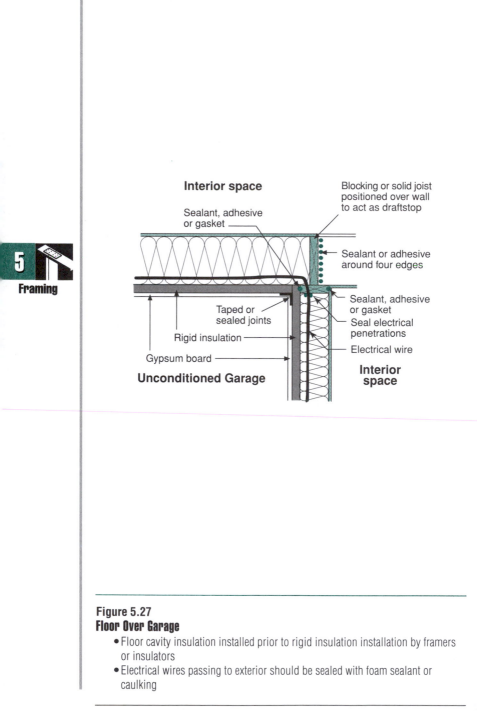

Interior space

Sealant, adhesive or gasket

Blocking or solid joist positioned over wall to act as draftstop

Sealant or adhesive around four edges

Taped or sealed joints

Rigid insulation

Gypsum board

Unconditioned Garage

Sealant, adhesive or gasket

Seal electrical penetrations

Electrical wire

Interior space

Figure 5.27
Floor Over Garage
- Floor cavity insulation installed prior to rigid insulation installation by framers or insulators
- Electrical wires passing to exterior should be sealed with foam sealant or caulking

Framing

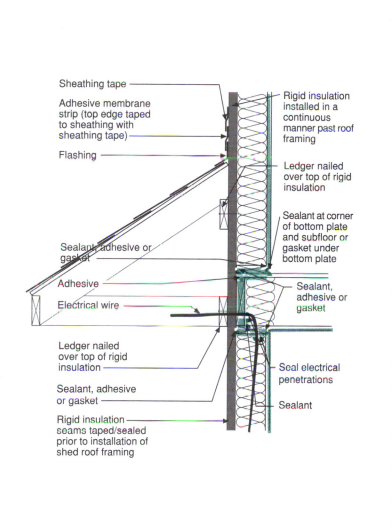

Sheathing tape

Adhesive membrane strip (top edge taped to sheathing with sheathing tape)

Flashing

Sealant, adhesive or gasket

Adhesive

Electrical wire

Ledger nailed over top of rigid insulation

Sealant, adhesive or gasket

Rigid insulation seams taped/sealed prior to installation of shed roof framing

Rigid insulation installed in a continuous manner past roof framing

Ledger nailed over top of rigid insulation

Sealant at corner of bottom plate and subfloor or gasket under bottom plate

Sealant, adhesive or gasket

Seal electrical penetrations

Sealant

Figure 5.28
Shed Roof

- Electrical wires passing to exterior should be sealed with foam sealant or caulking

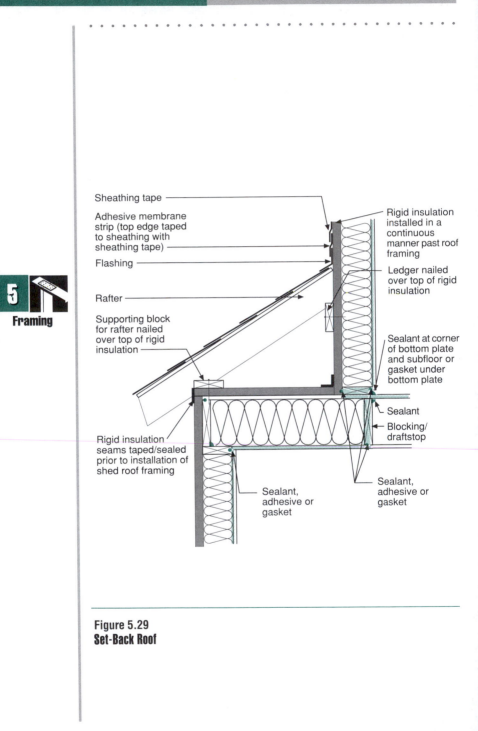

Sheathing tape

Adhesive membrane strip (top edge taped to sheathing with sheathing tape)

Flashing

Rafter

Supporting block for rafter nailed over top of rigid insulation

Rigid insulation seams taped/sealed prior to installation of shed roof framing

Rigid insulation installed in a continuous manner past roof framing

Ledger nailed over top of rigid insulation

Sealant at corner of bottom plate and subfloor or gasket under bottom plate

Sealant

Blocking/ draftstop

Sealant, adhesive or gasket

Sealant, adhesive or gasket

Figure 5.29
Set-Back Roof

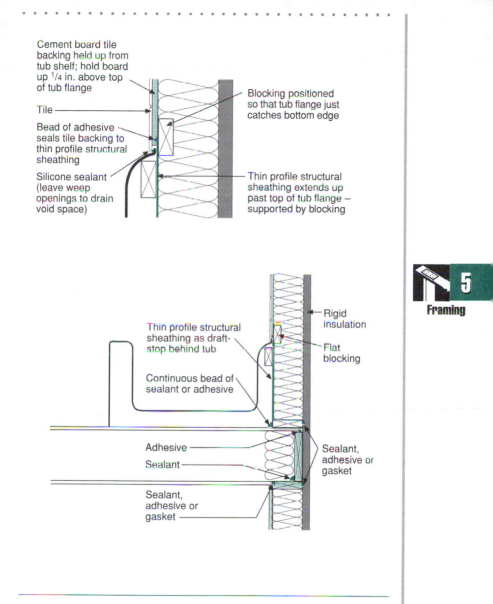

Cement board tile backing held up from tub shelf; hold board up $1/4$ in. above top of tub flange

Tile

Bead of adhesive seals tile backing to thin profile structural sheathing

Silicone sealant (leave weep openings to drain void space)

Blocking positioned so that tub flange just catches bottom edge

Thin profile structural sheathing extends up past top of tub flange – supported by blocking

Thin profile structural sheathing as draft-stop behind tub

Continuous bead of sealant or adhesive

Rigid insulation

Flat blocking

Adhesive

Sealant

Sealant, adhesive or gasket

Sealant, adhesive or gasket

5

Framing

Figure 5.30
Tub Framing – Section

- Flat blocking allows cavity insulation to be installed behind tub draftstop
- Cement board tile backing is recommended in place of "green board." However, if "green board" is used, it is critical that the factory edge be used at the top of the tub flange; alternatively, if a cut edge is used at this location it must be sealed with mastic to prevent wicking upwards.

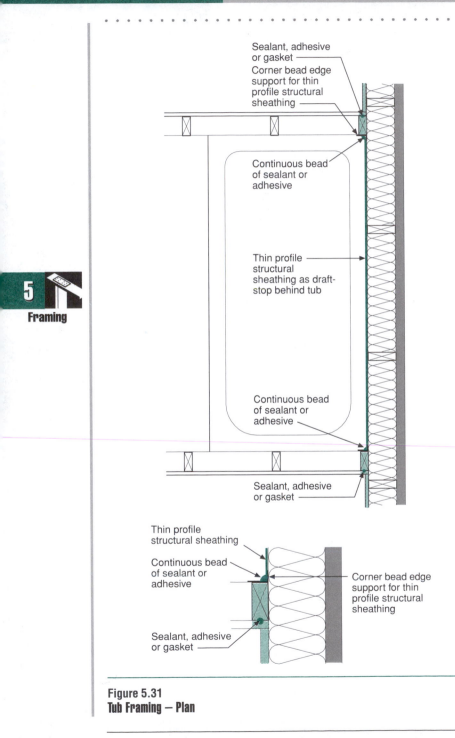

5

Framing

Figure 5.31
Tub Framing – Plan

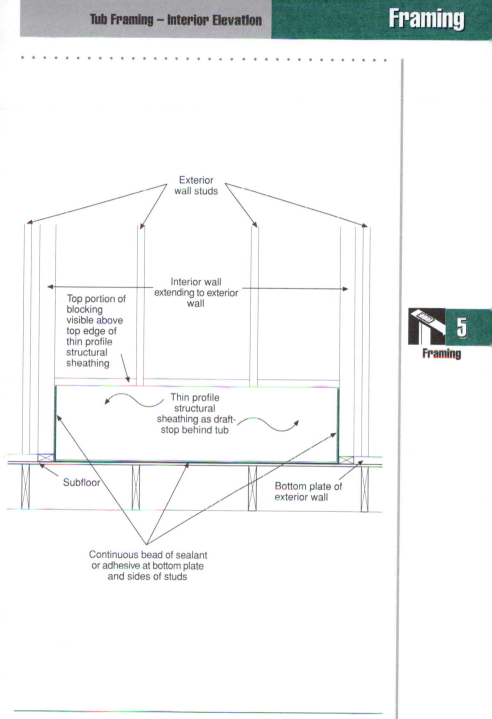

Exterior
wall studs

Interior wall
extending to exterior
wall

Top portion of
blocking
visible above
top edge of
thin profile
structural
sheathing

Thin profile
structural
sheathing as draft-
stop behind tub

Subfloor

Bottom plate of
exterior wall

Continuous bead of sealant
or adhesive at bottom plate
and sides of studs

5

Framing

Figure 5.32
Tub Framing – Interior Elevation

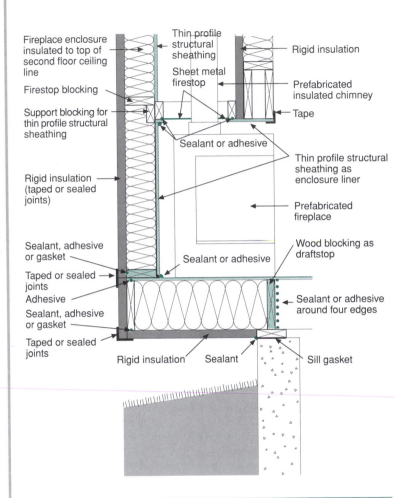

Figure 5.33
Fireplace Section

- Clearances around chimney to be determined by manufacturer's recommendations and local codes.
- Exterior combustion air with a damper should be provided to all fireboxes.
- Ideally, chimneys should be installed within the interior of the building envelope. Alternatively, chimney enclosures should be insulated full height to keep chimney flue pipes warm to ensure sufficient draft during the "die-down" stages of a fire. Insulated chimney flues are preferred.
- Use of sealed combustion, direct vent gas fireplaces eliminate the need for chimneys.

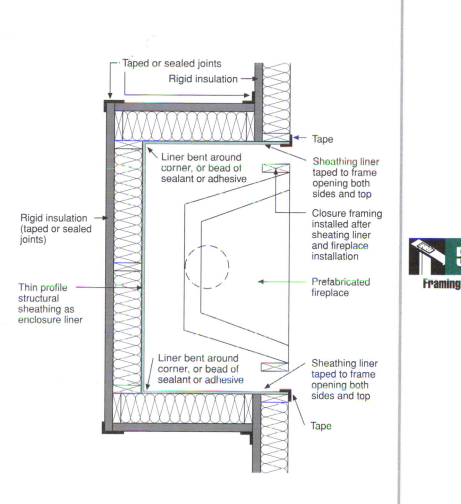

Taped or sealed joints

Rigid insulation

Tape

Liner bent around corner, or bead of sealant or adhesive

Sheathing liner taped to frame opening both sides and top

Rigid insulation (taped or sealed joints)

Closure framing installed after sheating liner and fireplace installation

Thin profile structural sheathing as enclosure liner

Prefabricated fireplace

Liner bent around corner, or bead of sealant or adhesive

Sheathing liner taped to frame opening both sides and top

Tape

5

Framing

Figure 5.34
Fireplace Plan

- Clearances around chimney to be determined by manufacturer's recommendations and local codes.
- Exterior combustion air with a damper should be provided to all fireboxes.
- Ideally, chimneys should be installed within the interior of the building envelope. Alternatively, chimney enclosures should be insulated full height to keep chimney flue pipes warm to ensure sufficient draft during the "die-down" stages of a fire. Insulated chimney flues are preferred.
- Use of sealed combustion, direct vent gas fireplaces eliminate the need for chimneys.

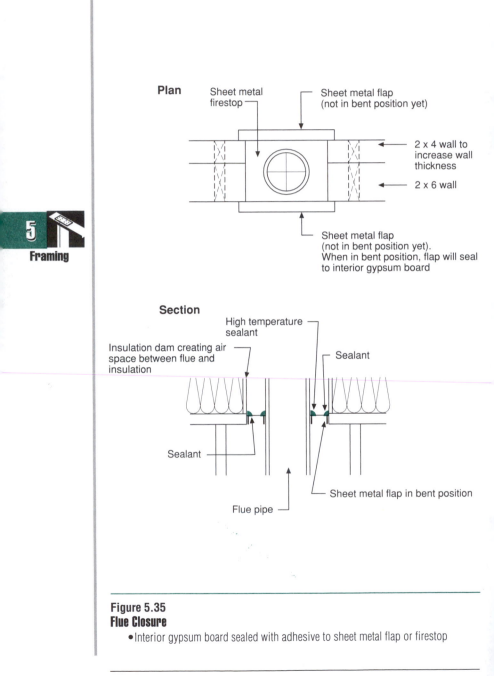

Plan

Sheet metal firestop

Sheet metal flap (not in bent position yet)

2 x 4 wall to increase wall thickness

2 x 6 wall

Sheet metal flap (not in bent position yet). When in bent position, flap will seal to interior gypsum board

Section

High temperature sealant

Insulation dam creating air space between flue and insulation

Sealant

Sealant

Sheet metal flap in bent position

Flue pipe

Figure 5.35
Flue Closure
- Interior gypsum board sealed with adhesive to sheet metal flap or firestop

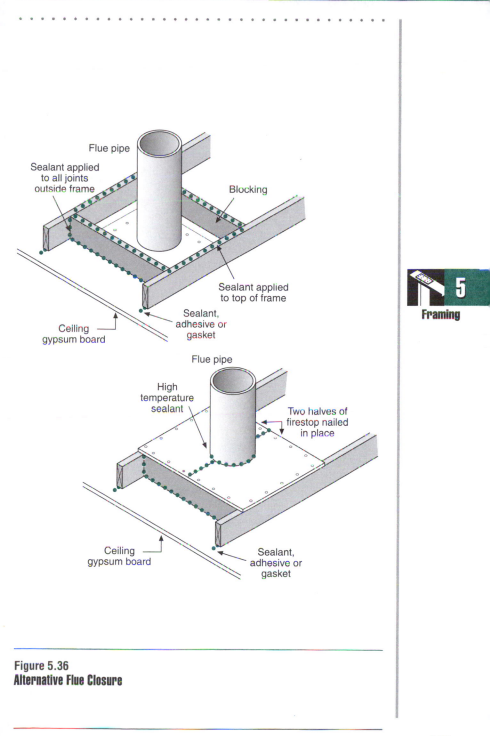

Flue pipe

Sealant applied
to all joints
outside frame

Blocking

Sealant applied
to top of frame

Ceiling
gypsum board

Sealant,
adhesive or
gasket

Flue pipe

High
temperature
sealant

Two halves of
firestop nailed
in place

Ceiling
gypsum board

Sealant,
adhesive or
gasket

5

Framing

Figure 5.36
Alternative Flue Closure

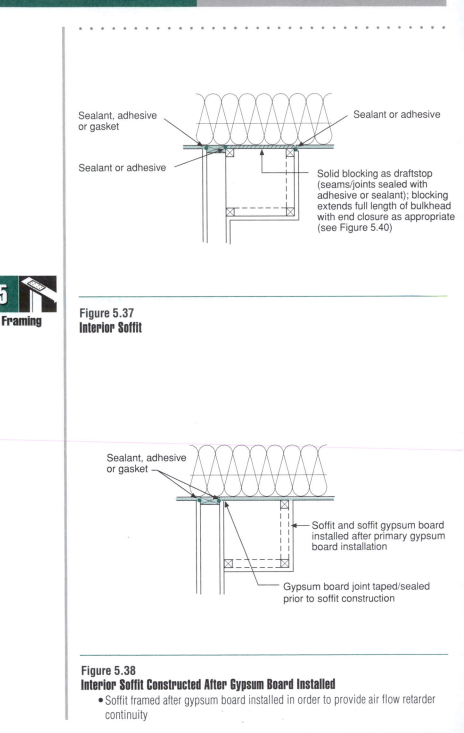

Sealant, adhesive or gasket

Sealant or adhesive

Sealant or adhesive

Solid blocking as draftstop (seams/joints sealed with adhesive or sealant); blocking extends full length of bulkhead with end closure as appropriate (see Figure 5.40)

5

Framing

Figure 5.37
Interior Soffit

Sealant, adhesive or gasket

Soffit and soffit gypsum board installed after primary gypsum board installation

Gypsum board joint taped/sealed prior to soffit construction

Figure 5.38
Interior Soffit Constructed After Gypsum Board Installed
- Soffit framed after gypsum board installed in order to provide air flow retarder continuity

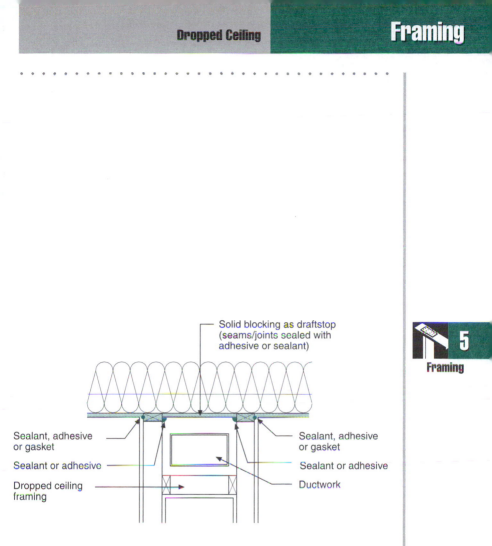

Solid blocking as draftstop
(seams/joints sealed with
adhesive or sealant)

5

Framing

Sealant, adhesive
or gasket

Sealant, adhesive
or gasket

Sealant or adhesive

Sealant or adhesive

Dropped ceiling
framing

Ductwork

Figure 5.39
Dropped Ceiling

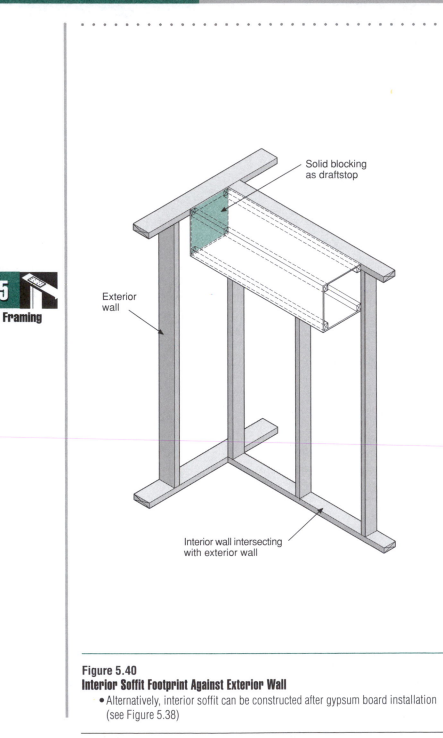

Solid blocking
as draftstop

Exterior
wall

Interior wall intersecting
with exterior wall

Figure 5.40
Interior Soffit Footprint Against Exterior Wall
- Alternatively, interior soffit can be constructed after gypsum board installation
(see Figure 5.38)

5
Framing

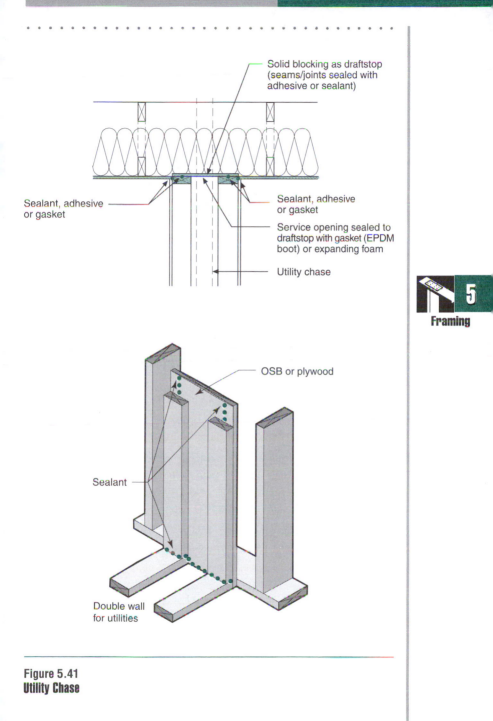

Solid blocking as draftstop
(seams/joints sealed with
adhesive or sealant)

Sealant, adhesive
or gasket

Sealant, adhesive
or gasket

Service opening sealed to
draftstop with gasket (EPDM
boot) or expanding foam

Utility chase

5

Framing

OSB or plywood

Sealant

Double wall
for utilities

Figure 5.41
Utility Chase

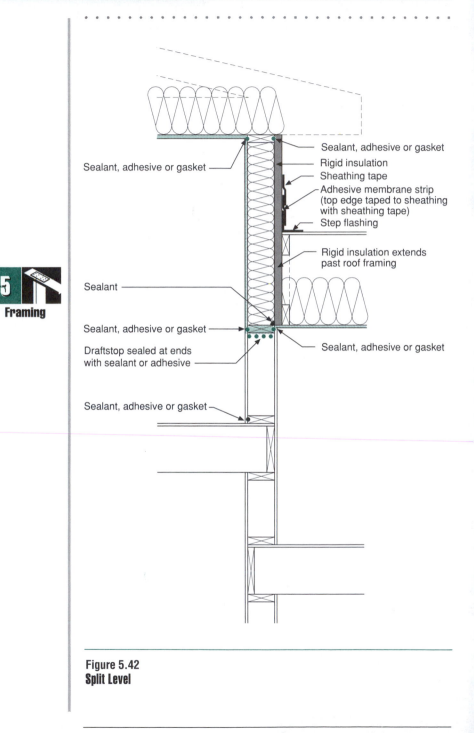

Sealant, adhesive or gasket

Sealant, adhesive or gasket

Rigid insulation

Sheathing tape

Adhesive membrane strip
(top edge taped to sheathing
with sheathing tape)

Step flashing

Rigid insulation extends
past roof framing

Sealant

Sealant, adhesive or gasket

Sealant, adhesive or gasket

Draftstop sealed at ends
with sealant or adhesive

Sealant, adhesive or gasket

Figure 5.42
Split Level

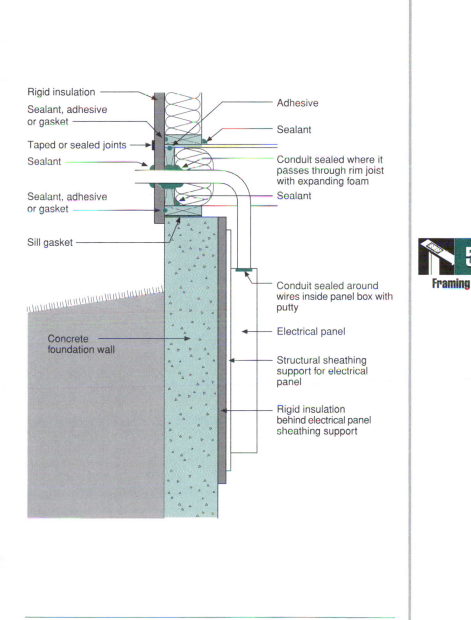

Rigid insulation

Sealant, adhesive or gasket

Taped or sealed joints

Sealant

Sealant, adhesive or gasket

Sill gasket

Concrete foundation wall

Adhesive

Sealant

Conduit sealed where it passes through rim joist with expanding foam

Sealant

Conduit sealed around wires inside panel box with putty

Electrical panel

Structural sheathing support for electrical panel

Rigid insulation behind electrical panel sheathing support

5

Framing

Figure 5.43
Electrical Panel
- Rigid insulation installed under electrical panel as a thermal break and to provide continuity for interior basement insulation

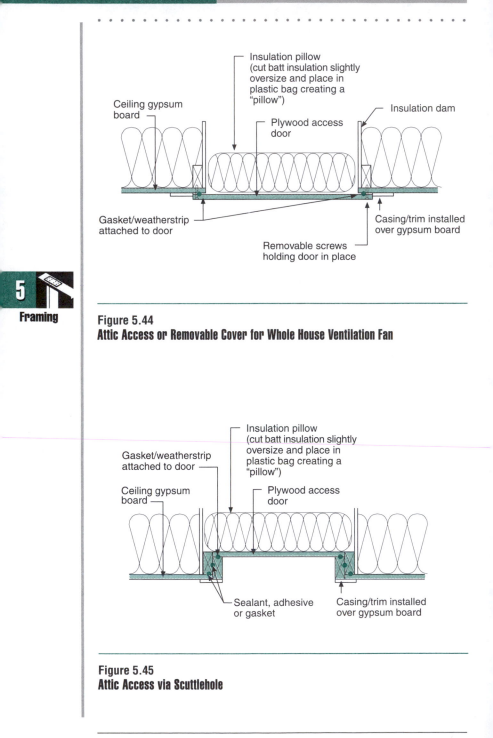

5

Figure 5.44
Attic Access or Removable Cover for Whole House Ventilation Fan

Figure 5.45
Attic Access via Scuttlehole

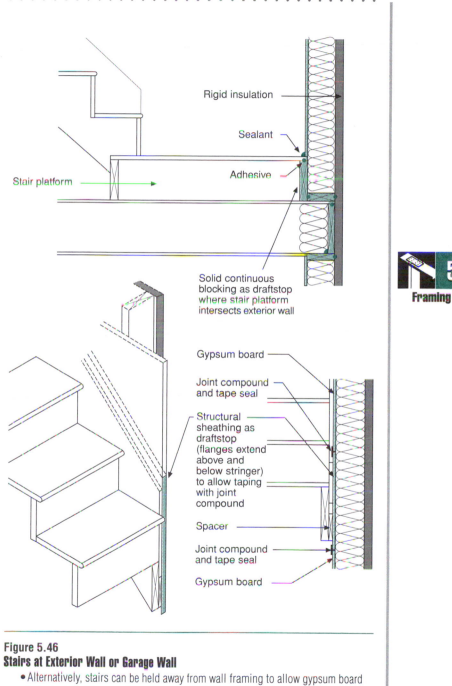

Rigid insulation

Sealant

Adhesive

Stair platform

Solid continuous blocking as draftstop where stair platform intersects exterior wall

Gypsum board

Joint compound and tape seal

Structural sheathing as draftstop (flanges extend above and below stringer) to allow taping with joint compound

Spacer

Joint compound and tape seal

Gypsum board

5

Framing

Figure 5.46
Stairs at Exterior Wall or Garage Wall
- Alternatively, stairs can be held away from wall framing to allow gypsum board to be installed in a continuous manner between stairs and wall framing

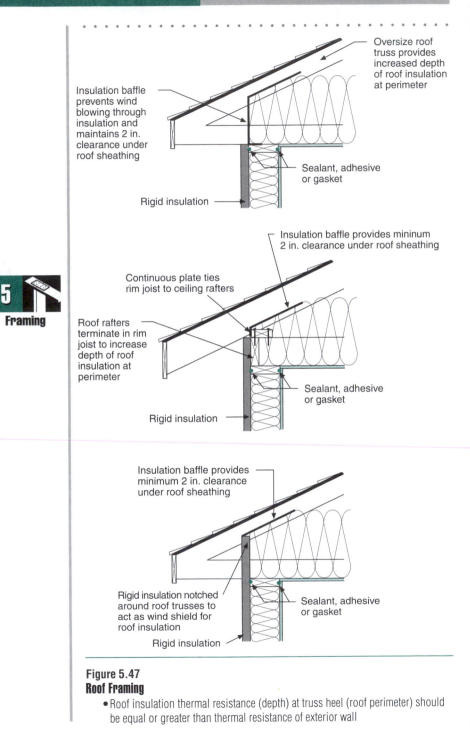

Insulation baffle prevents wind blowing through insulation and maintains 2 in. clearance under roof sheathing

Oversize roof truss provides increased depth of roof insulation at perimeter

Sealant, adhesive or gasket

Rigid insulation

Insulation baffle provides mininum 2 in. clearance under roof sheathing

Continuous plate ties rim joist to ceiling rafters

Roof rafters terminate in rim joist to increase depth of roof insulation at perimeter

Sealant, adhesive or gasket

Rigid insulation

Insulation baffle provides minimum 2 in. clearance under roof sheathing

Rigid insulation notched around roof trusses to act as wind shield for roof insulation

Sealant, adhesive or gasket

Rigid insulation

Figure 5.47
Roof Framing

- Roof insulation thermal resistance (depth) at truss heel (roof perimeter) should be equal or greater than thermal resistance of exterior wall

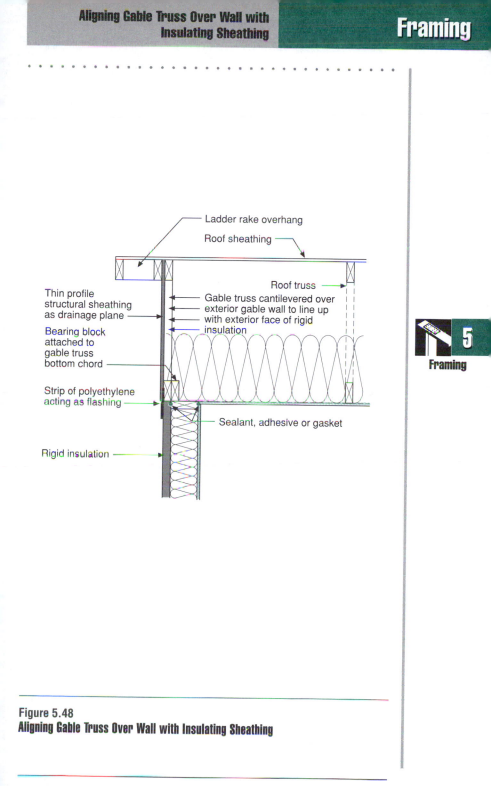

Ladder rake overhang

Roof sheathing

Roof truss

Thin profile structural sheathing as drainage plane

Gable truss cantilevered over exterior gable wall to line up with exterior face of rigid insulation

Bearing block attached to gable truss bottom chord

Strip of polyethylene acting as flashing

Sealant, adhesive or gasket

Rigid insulation

5

Framing

Figure 5.48
Aligning Gable Truss Over Wall with Insulating Sheathing

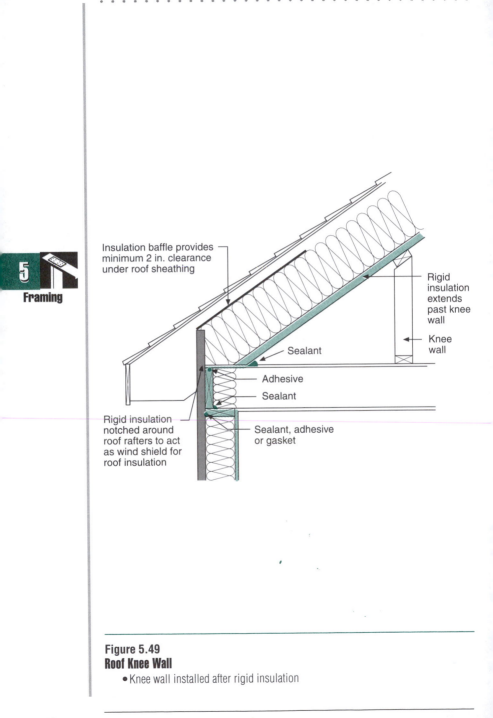

Insulation baffle provides minimum 2 in. clearance under roof sheathing

Rigid insulation extends past knee wall

Knee wall

Sealant

Adhesive

Sealant

Rigid insulation notched around roof rafters to act as wind shield for roof insulation

Sealant, adhesive or gasket

Figure 5.49
Roof Knee Wall
- Knee wall installed after rigid insulation

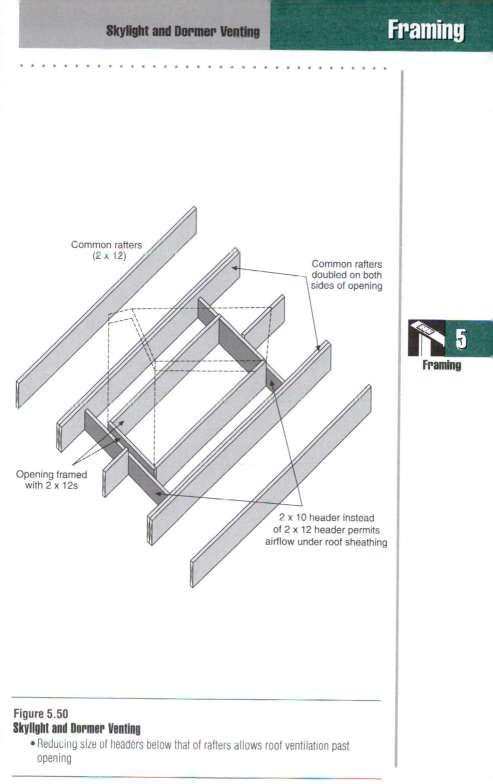

Common rafters
(2 x 12)

Common rafters
doubled on both
sides of opening

5

Framing

Opening framed
with 2 x 12s

2 x 10 header instead
of 2 x 12 header permits
airflow under roof sheathing

Figure 5.50
Skylight and Dormer Venting
- Reducing size of headers below that of rafters allows roof ventilation past
opening

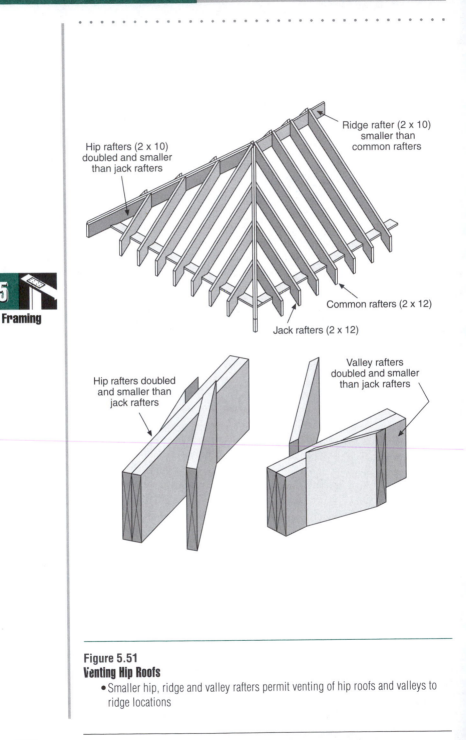

Figure 5.51
Venting Hip Roofs
- Smaller hip, ridge and valley rafters permit venting of hip roofs and valleys to ridge locations

5
Framing

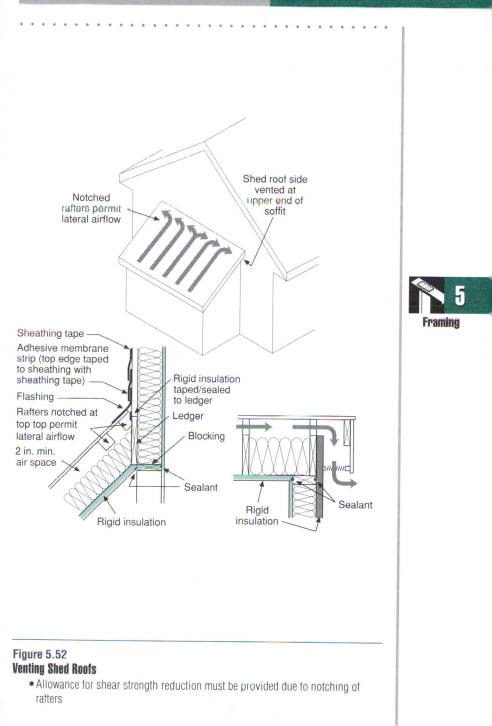

Figure 5.52
Venting Shed Roofs

- Allowance for shear strength reduction must be provided due to notching of rafters

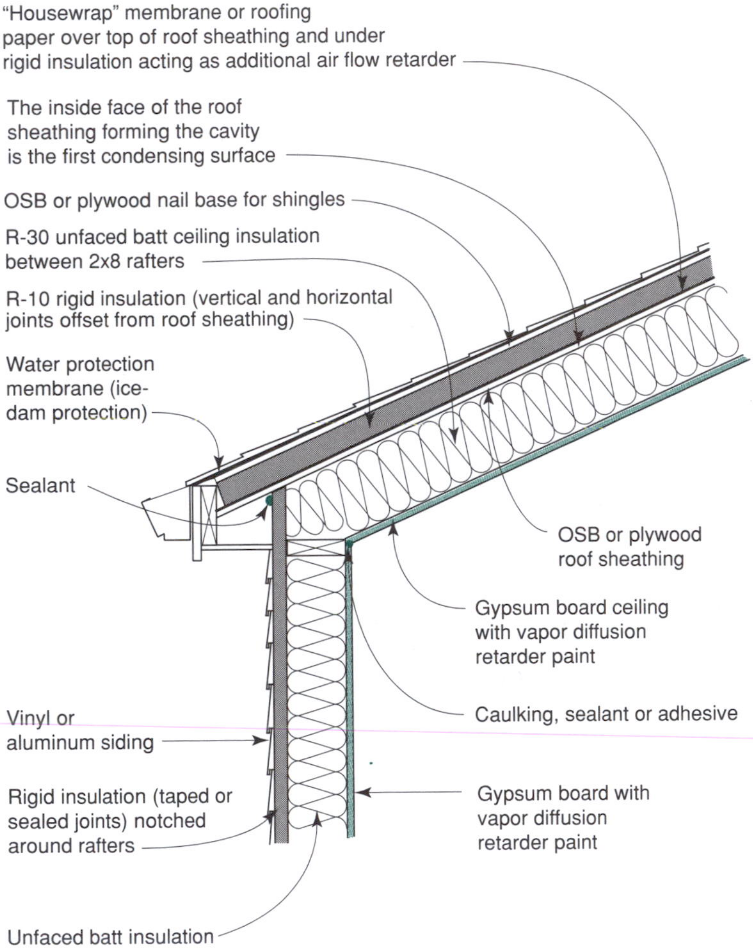

"Housewrap" membrane or roofing
paper over top of roof sheathing and under
rigid insulation acting as additional air flow retarder

The inside face of the roof
sheathing forming the cavity
is the first condensing surface

OSB or plywood nail base for shingles

R-30 unfaced batt ceiling insulation
between 2x8 rafters

R-10 rigid insulation (vertical and horizontal
joints offset from roof sheathing)

Water protection
membrane (ice-
dam protection)

Sealant

OSB or plywood
roof sheathing

Gypsum board ceiling
with vapor diffusion
retarder paint

Caulking, sealant or adhesive

Vinyl or
aluminum siding

Gypsum board with
vapor diffusion
retarder paint

Rigid insulation (taped or
sealed joints) notched
around rafters

Unfaced batt insulation

Figure 5.53
Hot Roof

- Rigid insulation raises dew point temperature of the first condensing surface
- This roof assembly is intended to be site constructed; prefabricated panels are
  not intended to be used in this assembly unless they are installed in a manner
  that provides air flow retarder continuity
- Offsetting joints in sheathing, rigid insulation and nail base creates air flow
  retarder continuity

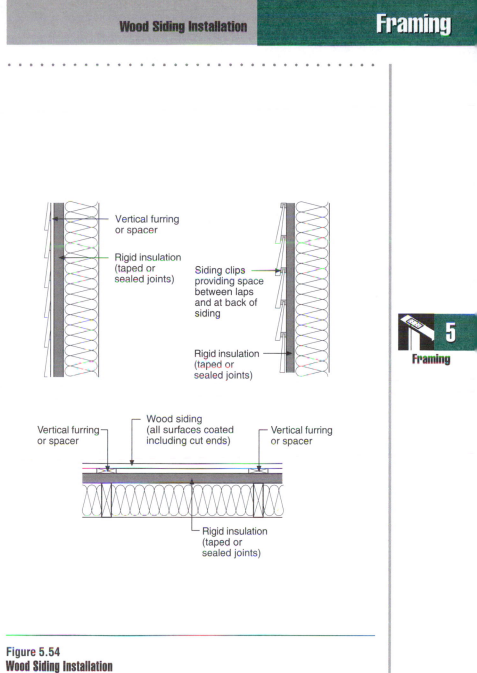

5

Framing

Framing

Figure 5.54
Wood Siding Installation
- Furring or spacer can be made by ripping strips of ³/₈" thick pressure treated plywood

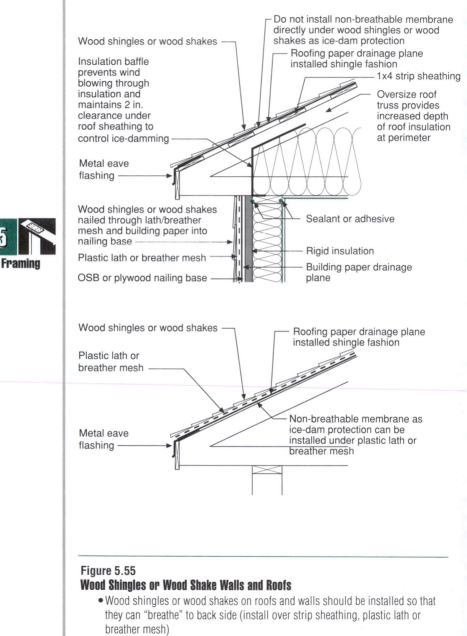

Do not install non-breathable membrane directly under wood shingles or wood shakes as ice-dam protection

Wood shingles or wood shakes

Roofing paper drainage plane installed shingle fashion

Insulation baffle prevents wind blowing through insulation and maintains 2 in. clearance under roof sheathing to control ice-damming

1x4 strip sheathing

Oversize roof truss provides increased depth of roof insulation at perimeter

Metal eave flashing

Wood shingles or wood shakes nailed through lath/breather mesh and building paper into nailing base

Sealant or adhesive

Plastic lath or breather mesh

Rigid insulation

OSB or plywood nailing base

Building paper drainage plane

Wood shingles or wood shakes

Roofing paper drainage plane installed shingle fashion

Plastic lath or breather mesh

Non-breathable membrane as ice-dam protection can be installed under plastic lath or breather mesh

Metal eave flashing

Figure 5.55
Wood Shingles or Wood Shake Walls and Roofs
- Wood shingles or wood shakes on roofs and walls should be installed so that they can "breathe" to back side (install over strip sheathing, plastic lath or breather mesh)
- Continuous soffit roof venting is more effective as ice-dam control than water protection membrane

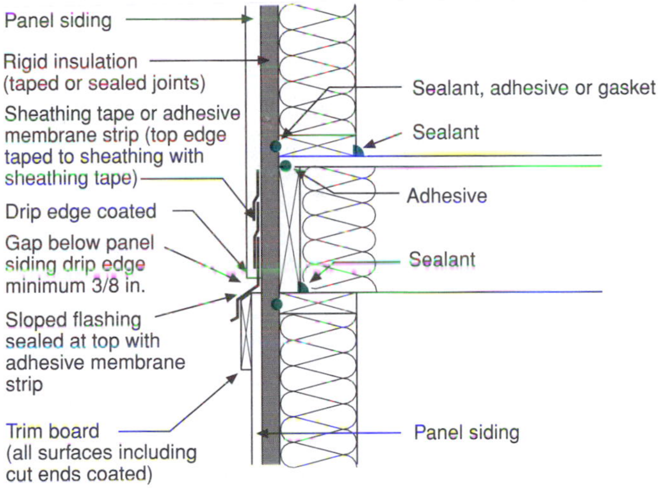

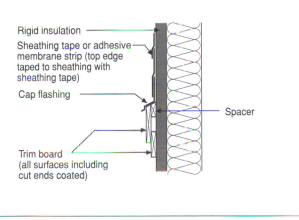

Panel siding

Rigid insulation
(taped or sealed joints)

Sheathing tape or adhesive
membrane strip (top edge
taped to sheathing with
sheathing tape)

Drip edge coated

Gap below panel
siding drip edge
minimum 3/8 in.

Sloped flashing
sealed at top with
adhesive membrane
strip

Trim board
(all surfaces including
cut ends coated)

Sealant, adhesive or gasket

Sealant

Adhesive

Sealant

Panel siding

Figure 5.56
Panel Trim

Rigid insulation

Sheathing tape or adhesive
membrane strip (top edge
taped to sheathing with
sheathing tape)

Cap flashing

Spacer

Trim board
(all surfaces including
cut ends coated)

Figure 5.57
Flashing Installed Over Padded Horizontal Trim

Framing

5

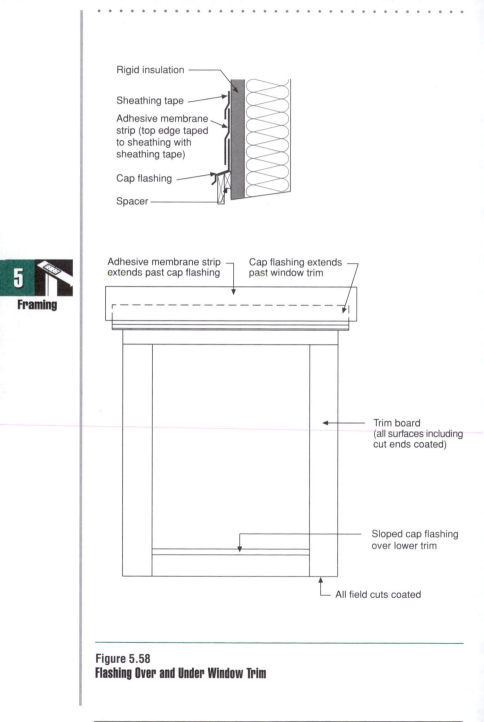

5

Framing

Rigid insulation

Sheathing tape

Adhesive membrane strip (top edge taped to sheathing with sheathing tape)

Cap flashing

Spacer

Adhesive membrane strip extends past cap flashing

Cap flashing extends past window trim

Trim board (all surfaces including cut ends coated)

Sloped cap flashing over lower trim

All field cuts coated

Figure 5.58
Flashing Over and Under Window Trim

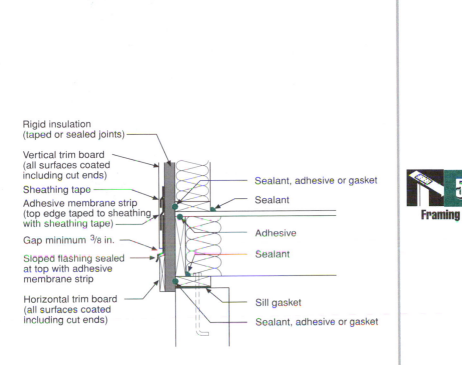

Rigid insulation
(taped or sealed joints)

Vertical trim board
(all surfaces coated
including cut ends)

Sheathing tape

Adhesive membrane strip
(top edge taped to sheathing
with sheathing tape)

Gap minimum $^3/_8$ in.

Sloped flashing sealed
at top with adhesive
membrane strip

Horizontal trim board
(all surfaces coated
including cut ends)

Sealant, adhesive or gasket

Sealant

Adhesive

Sealant

Sill gasket

Sealant, adhesive or gasket

5

Framing

Figure 5.59
Vertical Trim and Flashing Installed Over Horizontal Trim

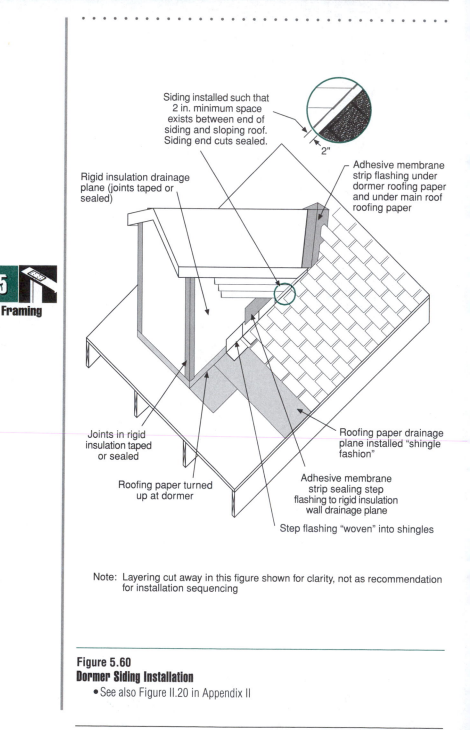

Siding installed such that 2 in. minimum space exists between end of siding and sloping roof. Siding end cuts sealed.

2"

Rigid insulation drainage plane (joints taped or sealed)

Adhesive membrane strip flashing under dormer roofing paper and under main roof roofing paper

Joints in rigid insulation taped or sealed

Roofing paper drainage plane installed "shingle fashion"

Roofing paper turned up at dormer

Adhesive membrane strip sealing step flashing to rigid insulation wall drainage plane

Step flashing "woven" into shingles

Note: Layering cut away in this figure shown for clarity, not as recommendation for installation sequencing

Figure 5.60
Dormer Siding Installation
• See also Figure II.20 in Appendix II

HVAC

Two heating and cooling approaches are common in residential construction: those with forced air and those without. Three controlled ventilation approaches are common in residential construction: exhaust systems, supply systems and balanced systems (Figures 6.1, 6.2, 6.3). When mixed and matched all these systems have to:

- heat when it's cold
- cool when it's hot
- humidify when it's dry
- dehumidify when it's wet
- bring in outside air
- distribute outside air
- exhaust strong pollutant point sources
- filter the air
- do all this when needed without noise, vibration, drafts and odors

Concerns

The first choice a builder makes is which type of energy source to use: combustion, electricity or the sun (if we split hairs, the sun is the source of all energy, except nuclear generated electricity - where we split atoms). Choosing between combustion, electricity or the sun requires the wisdom of Solomon, the intelligence of Newton and the wit of Wilde (the reason we have trouble with this choice is that none of these people are alive to actually ask). Wars have been fought over less. Sometimes no choice is possible. We will assume that some type of rational choice will be made.

If a combustion energy source is selected, combustion appliances should not be subject to backdrafting or spillage of combustion products. If electricity is selected, resistance heating should be avoided except in cases where building envelopes are so ultra-efficient that conventional thinking does not apply. In all cases, utilizing the sun for at least a partial contribution should be considered.

The second choice a builder makes is to use forced air or not. If air conditioning is desired (and by air conditioning we mean cooling), forced air systems usually become a necessity. If air conditioning is not a requirement, either forced air or radiant heating systems are options. Arguing the merits of forced air heating versus radiant heating systems can be compared to arguing about politics, religion or philosophy (we need to ask those dead guys again). Again, we will assume that some type of rational choice will be made.

Forced air heating and cooling systems involve an air handler with a cooling coil (such as a furnace with an A/C coil, a heat pump, or a fan-coil unit with an A/C coil and a heating coil coupled to a hot water heater). The air handler is typically connected to a supply and return ductwork system.

Radiant heating systems are those whose principal mode of heat transfer is radiation (sometimes coupled with convective heat transfer). Included are hot water or steam radiators, hot water baseboard heaters, electric baseboard convectors, radiant hot water floor piping, electric radiant panels and wood stoves. There are some radiant cooling systems available, but they are rare.

6

HVAC

The third choice a builder makes is how to ventilate. Ventilation? Remember, HVAC stands for heating, ventilating and air conditioning. Most builders seem to ignore the ventilating part. When builders think about it at all they seem to think they can handle ventilation by not building too tight and installing operable windows. Building leaky buildings or telling home owners to open windows somehow seems acceptable. How do you build a leaky building? Do you rely on lousy workmanship? Do you deliberately put holes in the building? How many holes do you put in and where do you put them? When you think about it, building leaky buildings doesn't make sense.

Why do we need to ventilate? We need to ventilate to protect building occupants and the building. Ventilation controls odors and airborne contaminants. Ventilation also can control interior moisture levels.

Ventilation Requirements

Building tight is right. Buildings can never be built too tight. However, they can be under ventilated.

The best way to go is to build a tight building envelope and install controlled ventilation using a fan or several fans which operate when people are present. Why the tight building envelope? Because before you can control air you must first enclose air. Why controlled ventilation? Because you don't want to over ventilate during cold weather and under ventilate during warm weather.

The stack effect provides the air pressure difference to drive air change in buildings. The stack effect occurs due to the density differences between heated and cooled air. Imagine a building as a hot air balloon too heavy to leave the ground with the stack effect causing the air at the top of the building to push upwards. A leaky building has a higher air change the colder it gets because the stack effect increases with increasing temperature difference. When it gets warmer, the stack effect disappears and you have almost no air change even though you may have a leaky building. If you have a hole, but no air pressure difference across the hole, no air flow occurs. Over ventilation during cold weather leads to comfort problems and heating bill complaints. Under ventilation during mild and warm weather leads to health complaints. Only a fan can provide a constant air change rate over the entire range of temperature differences. A controlled ventilation system chops off the peaks and fills in the valleys in air change rates as the seasons change.

How much air do you need? Somewhere between 10 cfm and 20 cfm per person when the building is occupied. If a building doesn't have strong interior pollutant sources, as low as 10 cfm per person will work. If a building has strong interior pollutant sources, not even 20 cfm per person will be enough. What are strong interior pollutant sources? Smokers. Damp basements. Pets that are not house trained. Unvented gas fireplaces or space heaters. Unusual hobby activities. Gas ovens and gas cooktops. Generally, if you keep the water out of a building, vent combustion appliances to the exterior, don't smoke, and don't have unusual habits or an uncommon lifestyle, 10 cfm per person will be just fine.

HVAC

How do you decide how many people live in a house? A good rule of thumb is to take the number of bedrooms and add 1. This assumes two people in the master bedroom and one person in each additional bedroom. At 10 cfm per person the following ventilation requirements result:

- two-bedroom house   30 cfm
- three-bedroom house   40 cfm
- four-bedroom house   50 cfm

Ventilation air should be provided when the building is occupied. Why ventilate when no one is in the building? Ventilation air should also be distributed (circulated) when the building is occupied.

Since most ventilation system airflow across the building envelope is in the 50 cfm or less range, the effect on building pressures is typically negligible. However, in general, exhaust ventilation systems have a slight depressurization effect on building enclosures; supply ventilation systems have a slight pressurization effect on building enclosures, and balanced ventilation systems have no effect on building air pressures. See Figures 6.4 through 6.10 for more detail on different types of ventilation systems. Note that the figures show side wall locations for the exit or entrance point of ventilation ducts because venting through the ceiling plane is not desirable due to the difficulties in air sealing fan housings and the effect of thermosiphoning on airflows.

Furnace fans or air handler fans should not run continuously unless efficient electric motors and blowers are used such as those with electrically commutated motors (ECM). These types of motors are typically found only on premium units. Alternatively, a flow controller can be used to cycle the blower several minutes every hour even if heating or cooling is not required. This will bring in outside air in cases where the outside air supply is provided by the blower (Figures 6.8 and 6.9).

Exhaust fans extracting less than 50 cfm will typically not increase radon ingress, soil gas ingress or backdrafting problems with fireplaces or wood stoves due to their negligible effect on building air pressures. Similarly, supply fans supplying less than 50 cfm will typically not increase wall and roof cavity interstitial moisture problems. Larger exhaust air flows may lead to unacceptably high negative air pressures (above 5 Pascals negative is considered unacceptable by some codes; 3 Pascals or less positive or negative is a recommended maximum allowable design metric for pressure differentials). Larger supply air flows may lead to moisture concerns in building envelope wall and roof assemblies that are not designed to dry towards the exterior or that do not have a provision to control condensing surface temperatures.

Combustion Appliances

Spillage or backdrafting of combustion appliances is unacceptable. If gas heating or a gas water heater is selected, the appliance must be power vented, sealed combustion or installed external to the conditioned space (garage). Traditional gas water heaters with draft hoods are prone to spillage and backdrafting. They should be avoided. Gas ovens, gas stoves or gas cooktops should only be installed with an exhaust range hood directly vented to the exterior. Wood-burning fireplaces or gas-burning fireplaces should be supplied with glass doors and exterior combustion air ducted to the firebox. Wood stoves should have a direct ducted supply of combustion air. Unvented (ventless) gas fireplaces or gas space heaters should never be installed. Sealed com-

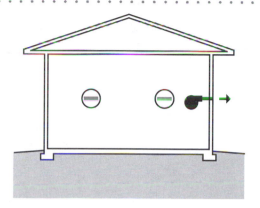

Figure 6.1
Exhaust Ventilation System

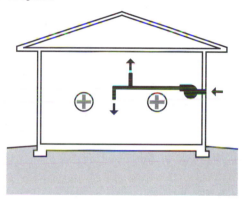

Figure 6.2
Supply Ventilation System

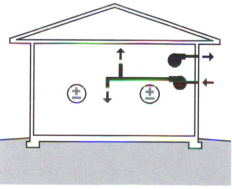

Figure 6.3
Balanced Ventilation System

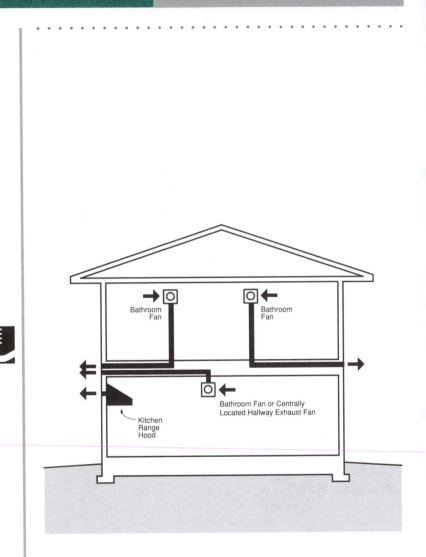

6

HVAC

Figure 6.4
Exhaust Ventilation System with Point Source Exhaust

- Individual exhaust fans pull interior air out of bathrooms. One of these fans is selected to also serve as the exhaust ventilation fan for the entire building with a run time based on time of occupancy. Alternatively, an additional centrally located (hallway) exhaust fan can be installed.
- Replacement air is drawn into bathrooms from hallways and bedrooms providing circulation and inducing controlled infiltration of outside air.
- Kitchen range hood provides point source exhaust as needed.

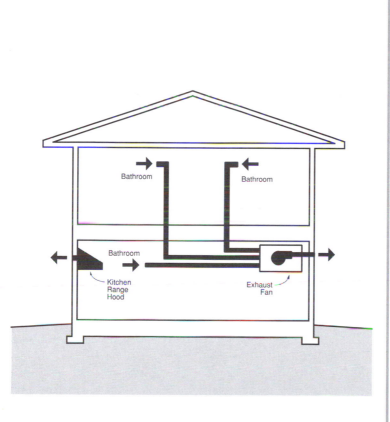

6

HVAC

Figure 6.5
Central Exhaust Ventilation System

- A single exhaust fan is ducted to individual bathrooms and pulls interior air out of bathrooms.
- Replacement air is drawn into bathrooms from hallways and bedrooms providing circulation and inducing controlled infiltration of outside air.
- Run time is based on time of occupancy.
- Individual bathroom fans are eliminated.
- Kitchen range hood provides point source exhaust as needed.

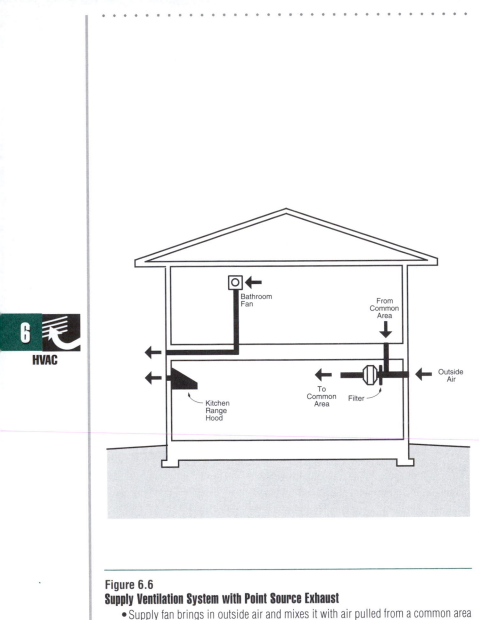

6

HVAC

Figure 6.6
Supply Ventilation System with Point Source Exhaust

- Supply fan brings in outside air and mixes it with air pulled from a common area (living room, hallway) to provide circulation and tempering prior to supplying to common area.
- Run time is based on time of occupancy.
- In supply ventilation systems, and with heat recovery ventilation, pre-filtration is recommended as debris can affect duct and fan performance reducing air supply.
- Kitchen range hood provides point source exhaust as needed.

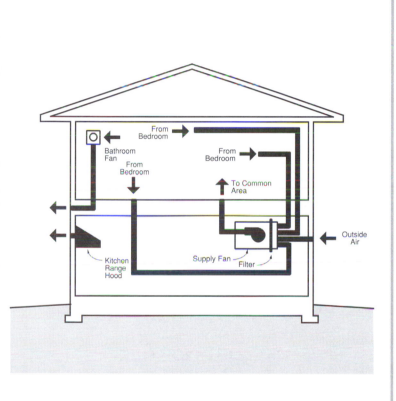

6

HVAC

Figure 6.7
Supply Ventilation System with Circulation and Point Source Exhaust
- Supply fan brings in outside air and mixes it with air pulled from bedrooms to provide circulation and tempering prior to supplying to common area.
- Run time is based on time of occupancy.
- In supply ventilation systems, and with heat recovery ventilation, pre-filtration is recommended as debris can affect duct and fan performance reducing air supply.
- Kitchen range hood provides point source exhaust as needed.

6

HVAC

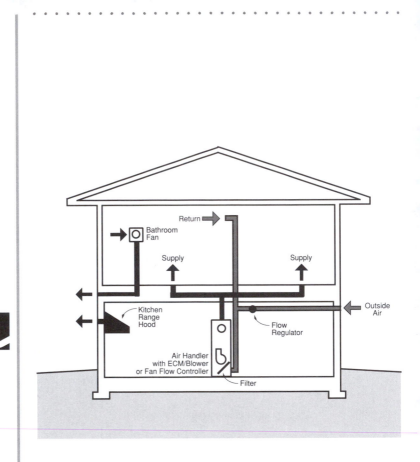

Figure 6.8
Supply Ventilation System Integrated with Heating and A/C

- Air handler with ECM/blower runs continuously (or operated based on time of occupancy by a flow controller) pulling outside air into the return system.
- A flow regulator provides fixed outside air supply quantities independent of air handler blower speed.
- House forced air duct system provides circulation and tempering.
- Point source exhaust is provided by individual bathroom fans and a kitchen range hood.
- In supply ventilation systems, and with heat recovery ventilation, pre-filtration is recommended as debris can affect duct and fan performance reducing air supply.
- Kitchen range hood provides point source exhaust as needed.
- Outside air duct should be insulated and positioned so that there is a fall/slope toward the outside to control any potential interior condensation. Avoid using long lengths of flex duct that may have a dip that could create a reservoir for condensation.

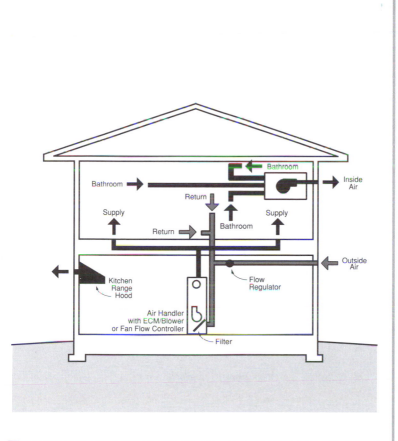

Figure 6.9

**Balanced Ventilation System Using a Supply Ventilation System Integrated with
Heating and A/C with a Stand Alone Central Exhaust**

- The supply system integrated with heating and A/C as in Figure 6.8.
- A central exhaust system is added as in Figure 6.5. Both systems are operated
  simultaneously.
- Run time is based on time of occupancy.
- Air handler with ECM/blower runs continuously (or operated based on time of
  occupancy by a flow controller) pulling outside air into the return system.
- In supply ventilation systems, and with heat recovery ventilation, pre-filtration is
  recommended as debris can affect duct and fan performance reducing air supply.
- Kitchen range hood provides point source exhaust as needed.
- Outside air duct should be insulated and positioned so that there is a fall/slope
  toward the outside to control any potential interior condensation. Avoid using
  long lengths of flex duct that may have a dip that could create a reservoir for
  condensation.

6

HVAC

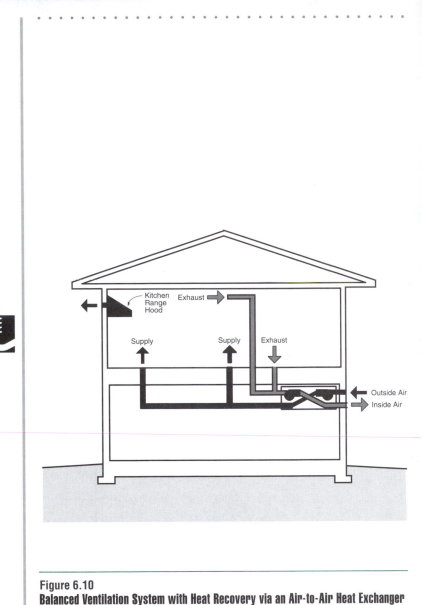

Figure 6.10
Balanced Ventilation System with Heat Recovery via an Air-to-Air Heat Exchanger
- The ventilation system has a separate duct system and is not integrated with the heating and A/C system.
- Run time is based on time of occupancy.
- Exhausts are typically from bathrooms and supplies are typically to bedrooms.
- In supply ventilation systems, and with heat recovery ventilation, pre-filtration is recommended as debris can affect duct and fan performance reducing air supply.

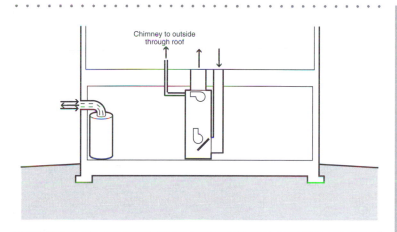

Figure 6.11
Direct Vented Gas Water Heater, Induced Draft Gas Furnace

- Water heater located against exterior wall; combustion air supplied directly to water heater from exterior via concentric duct; products of combustion exhausted directly to exterior also via concentric duct.
- Furnace flue gases exhausted to the exterior using a fan to induce draft; combustion air taken from the interior.

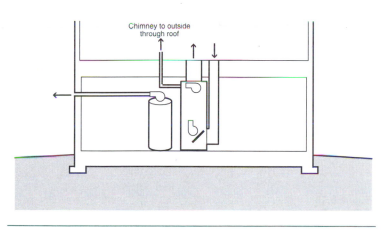

Figure 6.12
Power Vented Gas Water Heater, Induced Draft Gas Furnace

- Water heater flue gases exhausted to the exterior using a fan to maintain draft; combustion air taken from the interior.
- Furnace flue gases exhausted to the exterior using a fan to induce draft; combustion air taken from the interior.

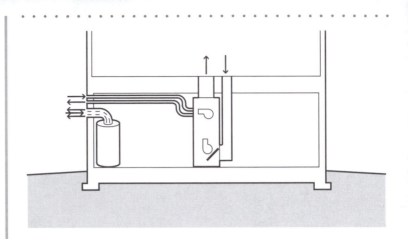

Figure 6.13
Direct Vented Gas Water Heater, Sealed Combustion Power Vented Furnace
- Water heater located against exterior wall; combustion air supplied directly to water heater from exterior via concentric duct; products of combustion exhausted directly to exterior also via concentric duct
- Furnace flue gases exhausted to the exterior using a fan; combustion air supplied directly to furnace from exterior via duct

6

HVAC

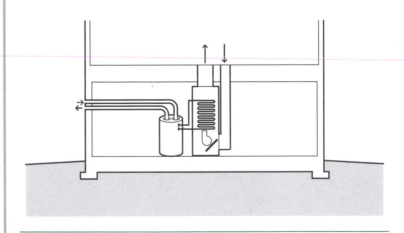

Figure 6.14
Sealed Combustion Power Vented Gas Water Heater
- Water heater flue gases exhausted to exterior using a fan; combustion air supplied directly to water heater from exterior via duct
- No furnace; heat provided by hot water pumped through a water-to-air heat exchanger (fan-coil)

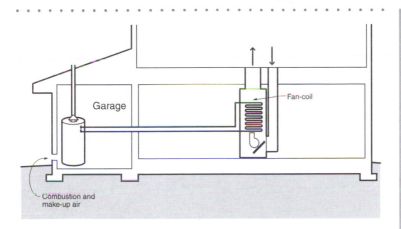

Figure 6.15
Traditional Gas Water Heater with a Water-to-Air Heat Exchanger
- Standard/traditional gas water heater with a draft hood is located out of the conditioned space in a garage supplied with combustion and make-up air
- No furnace; heat is provided by hot water pumped through a water-to-air heat exchanger (fan-coil)

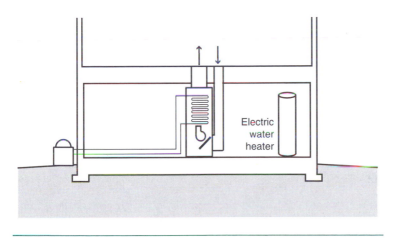

Figure 6.16
Air-to-Air Heat Pump
- Heating and cooling provided by an electrically driven heat pump with exterior air used as a heat source/sink

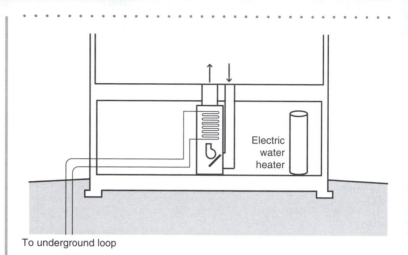

To underground loop

Figure 6.17
Ground Source Heat Pump
- Heating and cooling provided by an electrically driven heat pump with ground used as a heat source/sink

HVAC

bustion direct vent gas fireplaces are an acceptable alternative. Portable kerosene heaters should never be used indoors. Figures 6.11 through 6.15 describe several different systems for safely installing gas fired furnaces and gas hot water heaters.

Recirculating Fans

Recirculating range hoods and recirculating bathroom fans should be avoided due to health concerns. If recirculating range hood filters are not regularly replaced and units not regularly cleaned, they become a breeding ground for biologicals and a major source of odors.

Moisture Control

It is common to use dilution (controlled air change by an exhaust, supply or balanced ventilation system) during heating periods (cold, dry winter months) to limit/control interior moisture levels and the dehumidification characteristics of air conditioning (mechanical cooling) systems to reduce interior moisture levels during cooling periods (hot, sometimes humid, summer months).

However, dilution ventilation during the winter and air conditioning during the summer do not always control interior moisture levels in mixed climates. Furthermore, fall conditions, when neither heating nor cooling occurs and exterior humidity is high can lead to difficult interior moisture control problems. In some mixed climates, the winters are particularly mild, and the exterior air during the winter period is humid. Dilution (or air change) will not remove much moisture under these conditions since the incoming air is humid. Air change is still required to remove/dilute interior pollutants. However, air change will not remove moisture. Under such conditions, dehumidification is needed. A stand alone dehumidifier or a ventilating dehumidifier is recommended.

Similarly, under part load conditions in the summer or especially during the fall (when the outside air is humid, but at the same temperature or slightly higher than the interior) when the air conditioner does not operate much, dehumidification may also be needed. Air conditioning will only remove moisture from the interior air when the air conditioning system is cooling the interior air. If there is not much need (demand) for cooling, dehumidification by the air conditioning system will not occur. Again, under such conditions, dehumidification is needed and can be supplied by a stand alone dehumidifier or a ventilating dehumidifier.

Whole house fans are commonly used to provide cooling and comfort in mixed climates as an alternative to air conditioning (mechanical cooling) or as a means to displace some air conditioning load. Under many conditions when the exterior air is humid, whole house ventilation air conditioning energy savings are illusionary, due to the large latent (moisture) load typically associated with whole house ventilation.

HVAC

Whole house fans should only be operated when the exterior moisture levels (vapor pressure) are lower than the interior moisture levels (vapor pressure). Enthalpy (moisture) controllers are recommended for use with whole house fans. Interior surfaces and furnishing are hygroscopic (absorb/adsorb or hold moisture) and act as moisture reservoirs or batteries which are "charged" during cool, humid evenings when whole house fans are typically operated. This stored moisture is "discharged" into the interior air when the air conditioning system is finally turned on. The air conditioning system now has to work substantially harder/longer to cool the interior.

Whole house fans can also significantly depressurize building enclosures. Combustion appliance spillage, radon and soil gas infiltration concerns should be addressed by providing sub-slab ventilation systems, and by not using combustion appliances, such as fireplaces, when whole house fans are operating. Windows need to be open to provide make-up/relief air when whole house fans are operating.

System Sizing

Equipment should be sized correctly and return air flow paths should be planned. If a similar floor plan is constructed several times in a subdivision and sited with different orientations, heat gain and heat loss calculations should be done for each orientation. Equipment should be specifically selected for each orientation.

Incorrectly sized equipment can lead to operational and cost problems. Oversizing or undersizing heat pump systems with resistance strip back up (supplemental) heat, can alter the balance point and increase resistance strip heat use and, therefore, operating costs.

Oversizing air conditioning and heat pump systems can increase cycling losses, induce high wear and lead to loss of comfort control. During cooling periods, the dehumidification capabilities of the air conditioning system are used to control interior humidity. Oversizing or air conditioning equipment can lead to high interior humidity problems since oversized equipment will not run for extended periods of time, and, therefore, will dehumidify less than properly sized equipment.

HVAC

Incorrectly sized ductwork that is improperly laid out can also lead to operational and cost problems. Heat pumps and air conditioning systems lose significant operating system efficiencies if air flow volumes across coils are reduced as a result of improperly sized ducts, duct leakage or blockages due to layout, installation and a lack of servicing. If equipment is not located such that it is accessible, dirty coils and dirty filters will occur from a lack of servicing and result in a reduction of air flow. Too little air across the indoor coil can potentially lower the coil temperature to the point of ice formation and create serious damage.

Correct heat gain and heat loss determination is necessary in order to size and select equipment and systems. The Air Conditioning Contractors Association (ACCA) provides a recognized standard procedure in the publication, *Manual J*. Size of ductwork and distribution system calculation procedures are outlined in a second publication, *Manual D*.

Exterior coils of air conditioning and heat pump systems located under decks or adjacent vegetation will experience recirculation of air, resulting in greater operating losses.

Incorrectly selected controls, or controls adjusted incorrectly, can result in significant operational losses. Strip heat should be kept off with an outdoor thermostat that only allows operation when outdoor temperature is below the calculated balance point (e.g. 20 degrees Fahrenheit). Heat strips should be installed in banks 5 kW or less and each bank should have its own outdoor thermostat.

Thermostats for forced air systems are low voltage and typically have a built-in anticipation circuit, and a manual changeover switch or a heating/cooling lockout to prevent cross-cycling ("dueling") between heating and cooling modes. If thermostats have setback settings, they should have a ramped recovery or intelligent recovery feature that limits use of heat strips (supplementary heat) during the recovery period.

Incorrectly selected blower speeds can also result in significant operational losses. Blower speeds are usually different for cooling modes and heating modes. Approximately 425 to 450 cfm of air flow per ton of cooling is typically required over dry coils (400 cfm per ton for wet coils). Significantly less air flow is usually required through the same system during heating periods, and may vary further under different heating modes. Three to five blower speed settings are common and are usually set manually. Some units use two speeds, one for cooling, and a lower speed for heating. Blowers that automatically adjust for changes in duct resistance are also available.

Air Handlers and Ductwork

Furnaces, air handlers and ductwork should always be located within conditioned spaces with allowances for easy access to facilitate servicing, filter replacement, drain pan cleaning, and future replacement as technology improves. Necessary floor space or room within the conditioned space of the home should be provided. Furnaces, air handlers and ductwork should not be located in vented attics, vented crawl spaces or garages. Ductwork should not be located in exterior walls or in concrete floor slabs.

HVAC

Proper location of supply registers and return grilles is important to good system performance. The best approach involves either high supply registers and low return grilles, or low supply registers and high return grilles. Care should be taken to locate supply registers so that conditioned air is not blown directly on people occupying the space.

Locate air conditioning supply registers so that cold air is not blown directly across wall and ceiling surfaces. Improper placement can potentially chill these surfaces below dew point temperatures and cause mold growth and damage to interior finishes.

Figures 6.18 through 6.24 provide suggested conceptual ductwork layouts.

Ductwork, furnaces and air handlers should be sealed against air leakage. The only place air should be able to leave the supply duct system and the furnace or air handling unit is at the supply registers. The only place air should be able to enter the return duct system and the furnace or air handling unit is at the return grilles. A forced air system should be able to be pressure tested the way a plumber pressure tests a plumbing system for leaks. Builders don't accept leaky plumbing systems, they should not accept leaky duct systems.

Supply systems should be sealed with mastic in order to be airtight. All openings (except supply registers), penetrations, holes and cracks should be sealed with mastic or fiberglass mesh and mastic. Tape, especially duct tape, does not work and should not be used. Sealing of the supply system includes sealing the supply plenum, its attachment to the air handler or furnace, and the air handler or furnace itself. Joints, seams and openings on the air handler, furnace or ductwork near the air handler or furnace should be sealed with both fiberglass mesh and mastic due to greater local vibration and flexure (Figure 6.25).

Return systems should be "hard"-ducted and sealed with mastic in order to be airtight. Building cavities should never be used as return ducts. Stud bays or cavities should not be used for returns. Panned floor joists should not be used. Panning floor joists and using stud cavities as returns leads to leaky returns and the creation of negative pressure fields within interstitial spaces. Carpet dust marking at baseboards, odor problems, mold problems and pollutant transport problems typically occur when building cavities are used as return ducts.

The longitudinal seams and transverse joints in sheet metal ducts should be sealed (Figure 6.26). The inner liner of insulated plastic flex duct should be sealed where flex ducts are connected to other ducts, plenums, junction boxes and boots/registers (Figure 6.27).

In flex duct installation, the outer liner and insulation should be pulled back and the inner liner attached to the collar with a tie. Fiberglass mesh tape (fabric) should be installed over the inner liner and collar such that at least 1 inch of fiberglass mesh tape covers the exposed collar. Mastic is then applied over the fiberglass mesh tape. The insulation and the outer liner is then pulled back over the connection and sealed with a second tie (Figure 6.28). When flex ducts are used, care must be taken to prevent restricting air flow by "pinching" ducts.

Connections between grilles, registers and ducts at ceilings, floors or knee walls typically leak where the boot does not seal tightly to the grille or gypsum board. Air from the attic, basement, or crawl space can be drawn into the return. Leaks can also exist within the boot and where the ducts connect to the boot. Several examples of recommended installation are given in Figures 6.29 and 6.30.

If the gap between boots and gypsum board opening or subfloor openings is kept to less than $^3/_8$-inch, a bead of sealant or mastic may be used to seal the gap. Where gaps are larger than $^3/_8$-inch, fabric and mastic should both be used. The optimum approach is to keep the gaps to less than $^3/_8$-inch and use a bead of sealant. This requires careful coordination with the drywall contractor to make sure that the rough openings for the boots are cut no more than $^3/_8$-inch bigger than the actual boot size on all sides.

When a return plenum draws directly through a wall, the wall cavity may inadvertently become a duct. If the penetration through the wall is not blocked and sealed, return leaks can occur with air drawn from the wall cavity. The wall cavity should be isolated from the return (Figure 6.31).

Sometimes, return plenums leak through the floor. In floors with crawl spaces or basements, the plenum floor may not be airtight, allowing air from those zones to be drawn into the return.

In some slab homes, a 4-inch diameter chase pipe enters the plenum. The chase pipe carries the refrigerant lines, condensate piping, and control wiring which connect the indoor and outdoor units. This chase pipe is frequently unsealed, allowing unconditioned air or soil gases (radon, pesticides, herbicides, moisture) to be drawn into the return. Chases should never terminate inside the return air stream.

HVAC

Return plenums are sometimes formed by the enclosed space below the air handler support platform. This plenum will leak to adjacent walls and directly to the space in which it is located. A return plenum in an air handler closet may have no gypsum board separating it from an adjacent tub enclosure. As a result, air may be drawn from the attic. The adjacent walls often have plumbing and wiring in them that either comes from the attic, crawl space, garage, basement, outside, or some other interior space. Many of these platforms are lined with insulation or fibrous duct board because of fire codes and soundproofing requirements. This lining is not an air barrier and leakage will occur if the joints and penetrations are not sealed.

All supply registers should have clear access to a return grille in order to prevent the pressurization of bedrooms and the depressurization of common areas. Bedrooms should either have a direct-ducted return or a transfer grille. Undercutting of bedroom doors rarely works and should not be relied upon to relieve bedroom pressurization. A central "hard"-ducted return that is airtight and coupled with transfer grilles to relieve bedroom pressurization significantly outperforms a return system with leaky ducted returns in every room, stud bays used as return ducts and panned floor joists. See Figure 6.32 for an effective transfer grille detail.

In designs where ducts are unavoidably located in an unconditioned space, they should be sealed airtight and insulated. Penetrations of ductwork through building envelopes should also sealed (Figure 6.33).

The return side of the air handler or furnace also must be sealed, especially the filter access. The filter access should be sealed with aluminum tape to permit access. A roll of this tape should be left with the unit so that homeowners can re-tape access panels after filter replacement or other servicing.

Leakage can also occur at the connection between the air handler and the support platform. All sides of the air handler must be sealed to the support platform. Supply plenums also leak at the seams, particularly sleeved plenums. The preferred approach to seal supply and return plenums and gaps at air handlers is to use both fabric and mastic due to the greater flexure and vibration typically present near air handlers.

Large Exhaust Fans

Large exhaust fans and appliances such as whole house fans, attic ventilation fans, indoor grilles, clothes dryers, fireplaces and kitchen exhaust range hoods can significantly depressurize buildings.

When whole house fans are used for cooling, windows should be opened in order to relieve building pressures.

Attic ventilation fans should never be installed. A correctly constructed attic makes attic ventilation fans unnecessary. An incorrectly constructed attic should be repaired, not saddled with an energy wasting and problem-creating attic ventilation fan.

Kitchen exhaust range hoods should not be oversized. Anything larger than 100 cfm exhaust capacity for a kitchen exhaust range hood should be carefully integrated into the entire design of the building. Indoor barbecue grilles should only be installed with a provision for makeup air.

Clothes Dryers

Nothing practical can be done with the large unbalanced air flows created by clothes dryers except to reduce the impact of the negative effects by installing only sealed combustion, direct vent, induced draft or power vented combustion furnaces and water heaters. By the way, never duct a clothes dryer into the house. The resulting moisture load is more difficult to deal with than an intermittent pressure imbalance. Additionally, lint and other particulates are known to aggravate allergies and contribute to dustmarking on interior finishes.

It is important that dryers be located so that exhaust ducting can be installed within the manufacturers limitations on length and number of elbows. If long lengths and excessive elbows are unavoidable a booster fan designed for dryer use should be provided.

HVAC

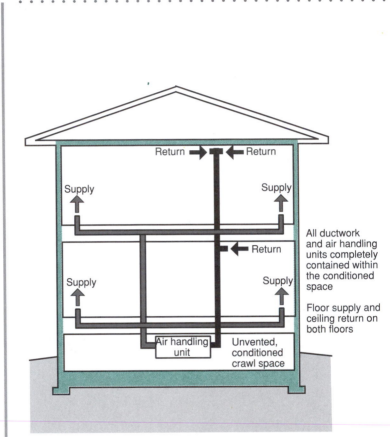

Note: Colored shading depicts the building's thermal barrier and pressure boundary. The thermal barrier and pressure boundary enclose the conditioned space.

6

HVAC

Figure 6.18
Unvented, Conditioned Crawl Space

- The air handling unit is located in an unvented, conditioned crawl space. The crawl space has a supply duct, but no return. A transfer grille is provided through the main floor to return air to the common area of the house and subsequently to the return grille on the main floor.
- Typical low efficiency gas appliances that are prone to spillage or backdrafting are not recommended in this type of application; heat pumps, heat pump water heaters or sealed combustion furnaces and water heaters should be used
- A hot water-to-air fan-coil in an air handling unit can be used to replace the gas furnace/gas burner. The fan-coil can be connected to a standard/traditional gas water heater with a draft hood located in the garage exterior to the building envelope "pressure boundary." Alternatively, the gas water heater can be sealed combustion (or power vented) and located within the conditioned space.

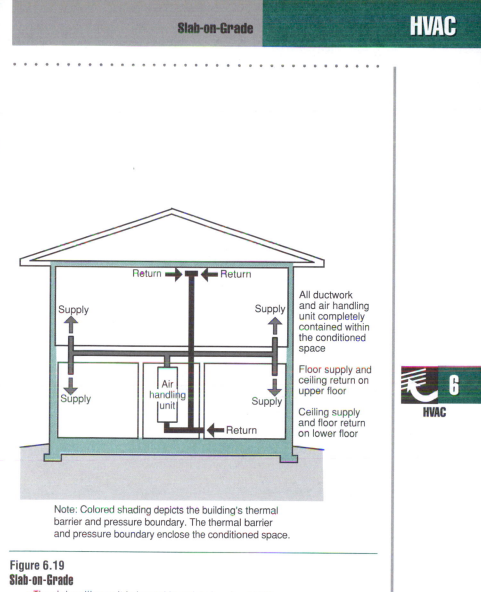

Return → ← Return

Supply ↑ Supply ↑

All ductwork and air handling unit completely contained within the conditioned space

Floor supply and ceiling return on upper floor

Supply ↓ Air handling unit Supply ↓

Ceiling supply and floor return on lower floor

← Return

Note: Colored shading depicts the building's thermal barrier and pressure boundary. The thermal barrier and pressure boundary enclose the conditioned space.

6

HVAC

Figure 6.19
Slab-on-Grade

- The air handling unit is located in an interior closet/utility room
- Typical low efficiency gas appliances that are prone to spillage or backdrafting are not recommended in this type of application; heat pumps, heat pump water heaters or sealed combustion furnaces and water heaters should be used
- A hot water-to-air fan-coil in an air handling unit can be used to replace the gas furnace/gas burner. The fan-coil can be connected to a standard/traditional gas water heater with a draft hood located in the garage exterior to the building envelope "pressure boundary." Alternatively, the gas water heater can be sealed combustion (or power vented) and located within the conditioned space.

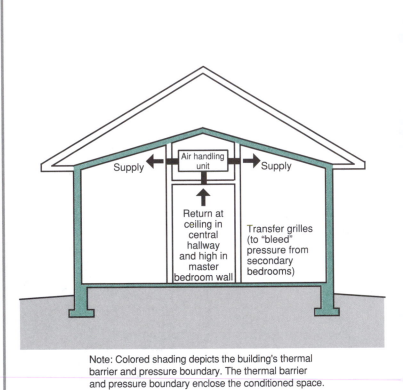

Note: Colored shading depicts the building's thermal barrier and pressure boundary. The thermal barrier and pressure boundary enclose the conditioned space.

Figure 6.20
Dropped Ceiling

- The air handling unit is located in an interior closet and the supply and return ductwork is located in a dropped hallway
- Transfer grilles "bleed" pressure from secondary bedrooms
- Ductwork does not have to extend to building perimeters when thermally efficient windows (low E, spectrally selective) and thermally efficient (well insulated 2x6 frame walls with 1" of insulating sheathing) building envelope construction is used; throw type of registers should be selected
- Typical low efficiency gas appliances that are prone to spillage or backdrafting are not recommended in this type of application; heat pumps, heat pump water heaters or sealed combustion furnaces and water heaters should be used
- A hot water-to-air fan-coil in an air handling unit can be used to replace the gas furnace/gas burner. The fan-coil can be connected to a standard/traditional gas water heater with a draft hood located in the garage exterior to the building envelope "pressure boundary." Alternatively, the gas water heater can be sealed combustion (or power vented) and located within the conditioned space.

6

HVAC

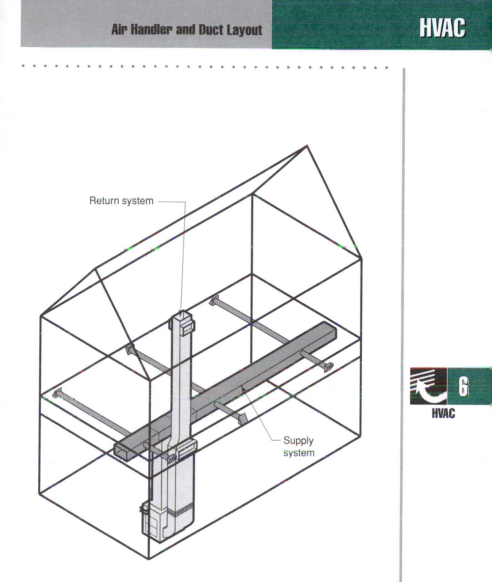

Return system

Supply system

Figure 6.21
Air Handler and Duct Layout
- Air handler centrally located to minimize duct runs
- No ductwork in exterior walls or attic
- Return high in hallway of upper floor
- Return low in hallway of main level
- Only fully "hard"-ducted returns when connected directly to air handler; no panned floor joist returns; no stud cavity returns
- Either return ducts in bedrooms or transfer grilles

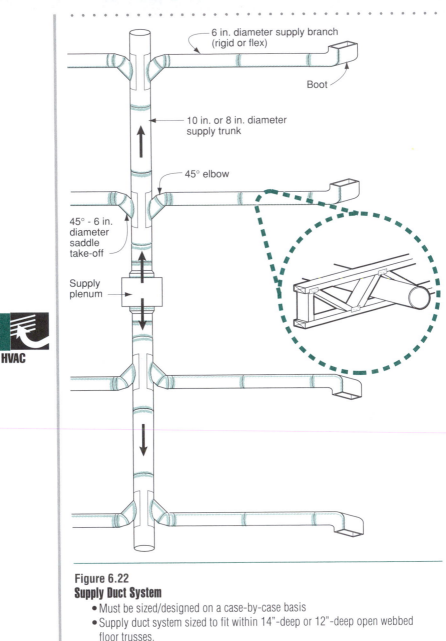

6

HVAC

Figure 6.22
Supply Duct System

- Must be sized/designed on a case-by-case basis
- Supply duct system sized to fit within 14"-deep or 12"-deep open webbed floor trusses.
- Trunk ducts can be 10" diameter for 14"-deep floor trusses or 8" diameter for 12"-deep floor trusses depending on air flows and heat losses
- Branch supply ducts can be insulated flex
- Rounded ducts, 45° take-offs and 45° elbows reduce air flow resistance so ducts can be made smaller to fit in floor system

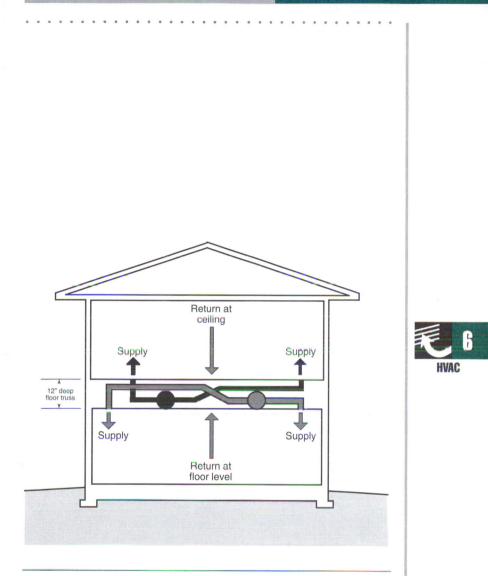

6

HVAC

Figure 6.23
Floor Truss Supply Duct System

- A 6" diameter flex duct can cross over an 8" diameter flex-duct within a 12"-deep floor cavity. At the point of crossing both ducts assume an oval cross section without a resultant meaningful/significant pressure drop.
- Flows between first and second floor can be balanced
- A return is located at the top of the second floor at the ceiling level
- A return is also located at the first floor at the floor level
- Transfer grilles are installed between bedrooms and hallways
- Air handler can be located in dropped ceiling below second floor framing

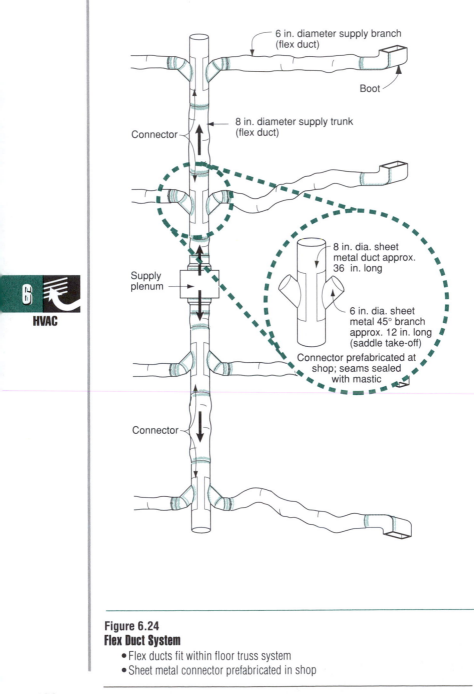

HVAC

6

Figure 6.24
Flex Duct System

- Flex ducts fit within floor truss system
- Sheet metal connector prefabricated in shop

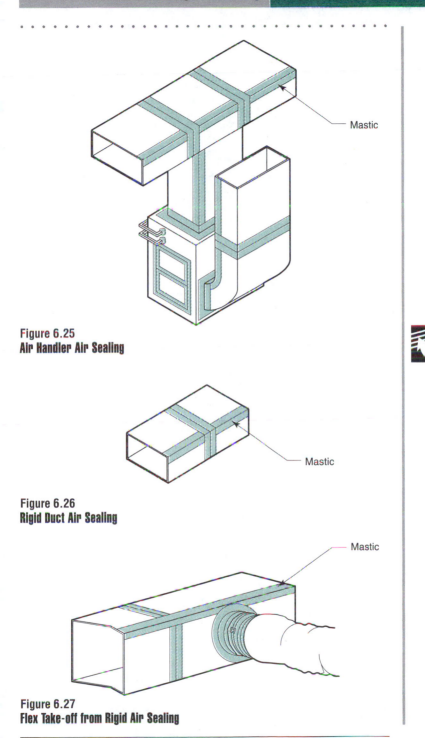

Mastic

Figure 6.25
Air Handler Air Sealing

Mastic

Figure 6.26
Rigid Duct Air Sealing

Mastic

Figure 6.27
Flex Take-off from Rigid Air Sealing

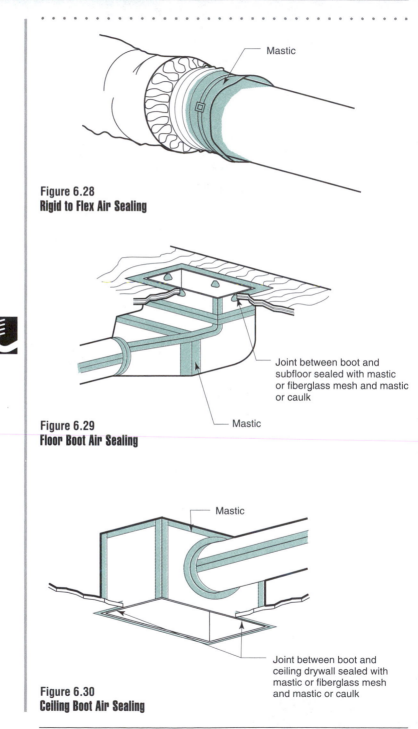

Figure 6.28
Rigid to Flex Air Sealing

Mastic

Figure 6.29
Floor Boot Air Sealing

Joint between boot and subfloor sealed with mastic or fiberglass mesh and mastic or caulk

Mastic

Figure 6.30
Ceiling Boot Air Sealing

Mastic

Joint between boot and ceiling drywall sealed with mastic or fiberglass mesh and mastic or caulk

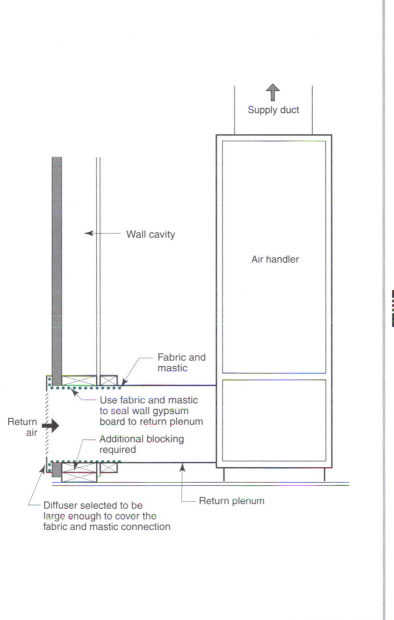

Supply duct

Wall cavity

Air handler

Fabric and mastic

Use fabric and mastic to seal wall gypsum board to return plenum

Return air

Additional blocking required

Diffuser selected to be large enough to cover the fabric and mastic connection

Return plenum

6

HVAC

Figure 6.31
Sealing Return Plenums

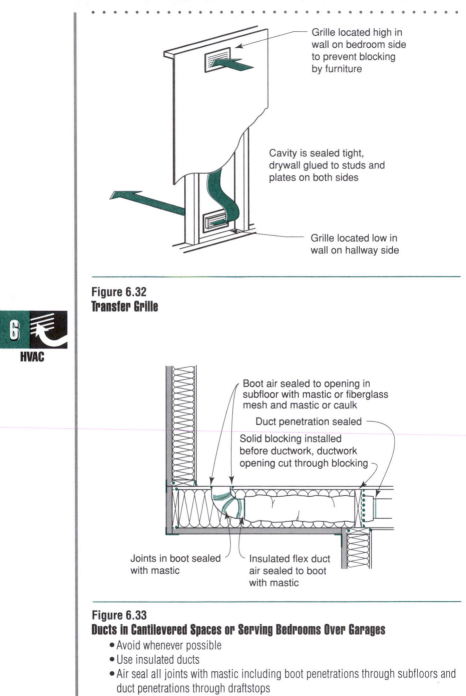

Grille located high in wall on bedroom side to prevent blocking by furniture

Cavity is sealed tight, drywall glued to studs and plates on both sides

Grille located low in wall on hallway side

Figure 6.32
Transfer Grille

6

HVAC

Boot air sealed to opening in subfloor with mastic or fiberglass mesh and mastic or caulk

Duct penetration sealed

Solid blocking installed before ductwork, ductwork opening cut through blocking

Joints in boot sealed with mastic

Insulated flex duct air sealed to boot with mastic

Figure 6.33
Ducts in Cantilevered Spaces or Serving Bedrooms Over Garages
- Avoid whenever possible
- Use insulated ducts
- Air seal all joints with mastic including boot penetrations through subfloors and duct penetrations through draftstops
- Consider spraying boot exterior with foam insulation

7

Plumbing

Plumbing systems have to:

- supply cold water

- supply hot water

- remove gray water and solid wastes

- not leak water, odors or air

Concerns

Plumbing system penetrations can be a major source of air leakage. Don't put plumbing in outside walls. Let us repeat that for those that may not get it. Don't put plumbing in outside walls. Holes in outside walls cause drafts; outside walls get cold. Pipes freeze. Owners get annoyed. Get it?

Plumbing penetrations through rim joists should be sealed with expanding foam or caulk. Vent stacks penetrating into attics should be sealed with flexible seals to handle expansion of pipes. See Figures 7.1 through 7.3 for details.

Tubs and Shower Stalls

Tubs, shower stalls, and one-piece tub-shower enclosures installed on exterior walls can be one of the single largest sources of air leakage across a building envelope. It is essential that rigid draftstopping material is installed prior to tub and shower stall installation. With one-piece tub-shower enclosures, the entire height of the interior surface of the exterior wall should be insulated and sheathed prior to tub-shower enclosure installation. See Figures 5.28 through 5.30 for details.

Water Consumption

Low-flow toilets and shower heads should be installed to minimize water consumption. Pressure balanced shower controls should be used to reduce the dangers of scalding.

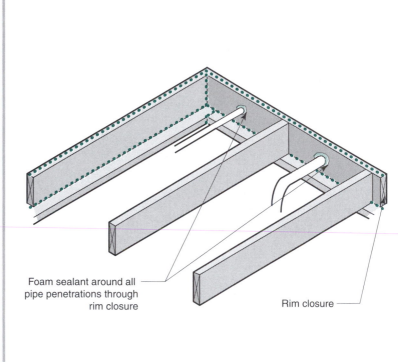

Foam sealant around all pipe penetrations through rim closure

Rim closure

Figure 7.1
Rim Penetrations

- Sealant should be installed at the rim closure itself, usually from the inside; sealing at the exterior siding (i.e. aluminum or vinyl siding) is ineffective

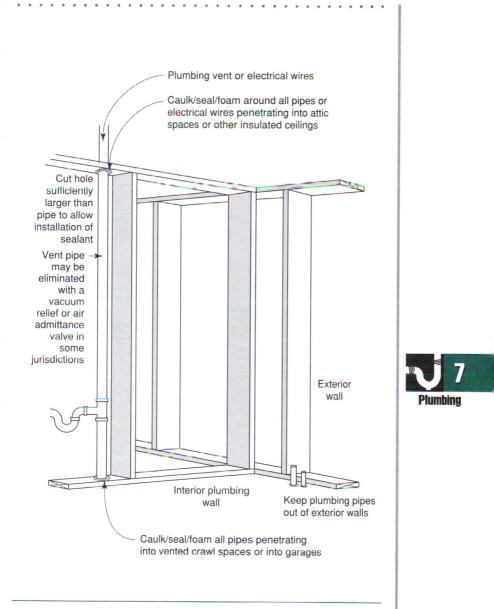

Plumbing vent or electrical wires

Caulk/seal/foam around all pipes or electrical wires penetrating into attic spaces or other insulated ceilings

Cut hole sufficiently larger than pipe to allow installation of sealant

Vent pipe may be eliminated with a vacuum relief or air admittance valve in some jurisdictions

Exterior wall

Plumbing

Interior plumbing wall

Keep plumbing pipes out of exterior walls

Caulk/seal/foam all pipes penetrating into vented crawl spaces or into garages

Figure 7.2
Locating Plumbing Pipes

- Where connections are made to toilets and pedestal sinks and exposed piping, chrome plated or other surface finished materials should be considered for aesthetic reasons
- Sealants should be flexible or non-hardening

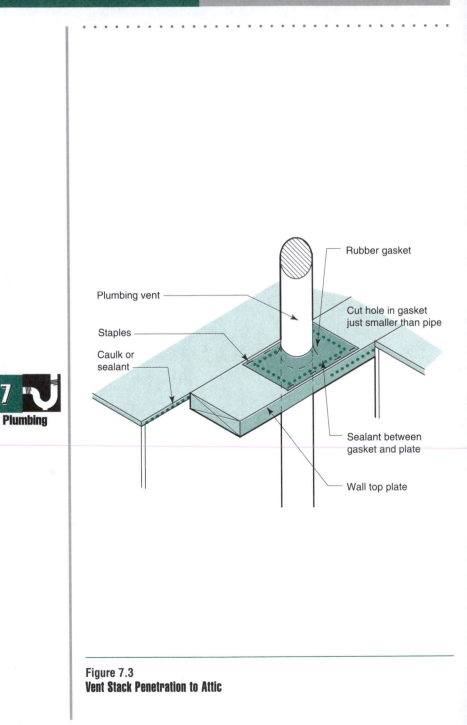

Figure 7.3
Vent Stack Penetration to Attic

Electrical

Electrical systems have to:

- supply electricity

- supply communication and control signals

- not leak air

Concerns

Electrical system penetrations through the building envelope can be a major source of air leakage.

Airtight outlet boxes should be installed in exterior walls and insulated ceilings. Specialized boxes are available. Alternatively, sealants can be used to seal penetrations in standard outlet boxes. See Figure 8.2 for details.

Electrical

Electrical penetrations through rim joists should be sealed with expanding foam or caulk. Wires penetrating into attics, and through top and bottom plates in exterior walls should be sealed with expanding foam or caulk. Air can also leak through service penetrations in studs where interior walls intersect exterior walls. These penetrations should also be sealed (Figure 8.3).

Recessed light fixtures in insulated ceilings should be insulation cover (IC) rated fixtures which are airtight and can be covered with insulation (Figure 8.4). Recessed light fixtures installed in dropped ceilings or soffits need to be draftstopped (Figure 8.5).

Where electrical panels are installed on exterior walls, air sealing of all penetrations is necessary (Figure 8.1).

Wires should be located along plates or against studs rather than through the center of insulated cavities to minimize insulation compression where batt insulation is used (Figure 8.4).

Lighting fixtures, locations and approaches should be selected in conjunction with daylighting design. Energy efficient lighting fixtures, bulbs and controls should be specified.

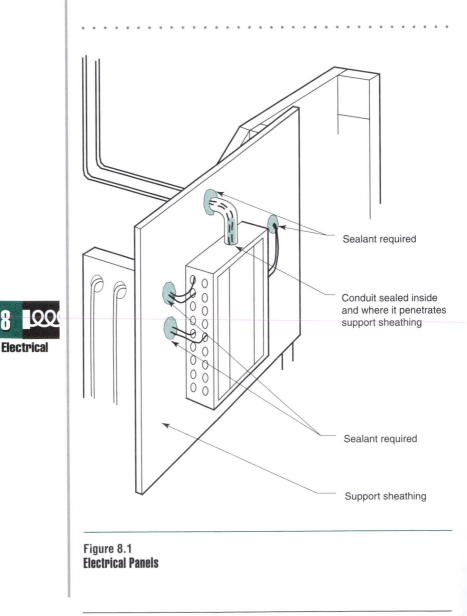

Sealant required

Conduit sealed inside and where it penetrates support sheathing

Sealant required

Support sheathing

Figure 8.1
Electrical Panels

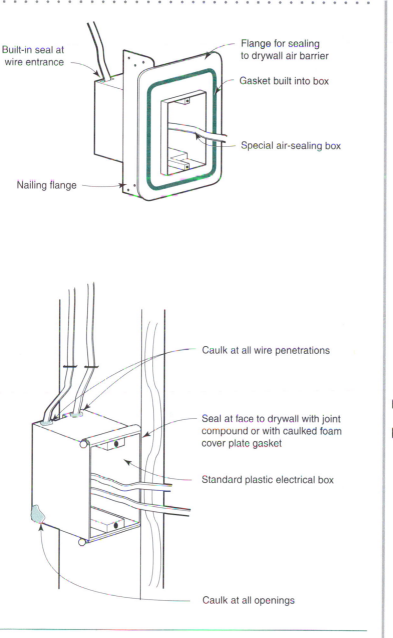

Built-in seal at
wire entrance

Flange for sealing
to drywall air barrier

Gasket built into box

Special air-sealing box

Nailing flange

Caulk at all wire penetrations

Seal at face to drywall with joint
compound or with caulked foam
cover plate gasket

Standard plastic electrical box

Caulk at all openings

Figure 8.2
Electrical Boxes

8

Electrical

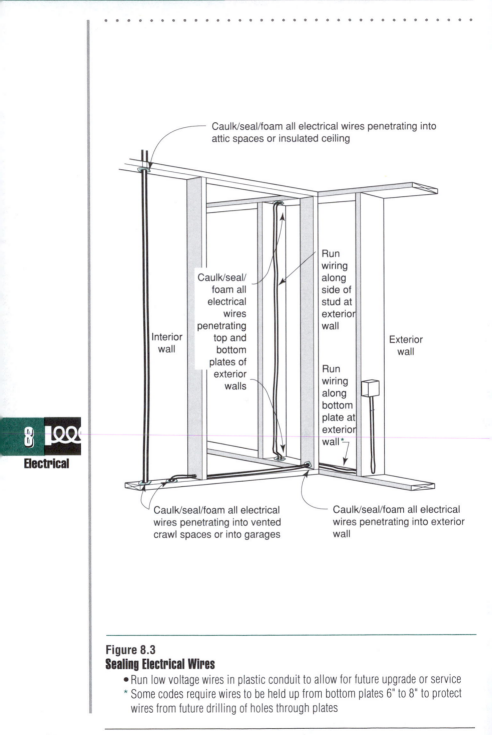

Caulk/seal/foam all electrical wires penetrating into attic spaces or insulated ceiling

Run wiring along side of stud at exterior wall

Caulk/seal/ foam all electrical wires penetrating top and bottom plates of exterior walls

Interior wall

Exterior wall

Run wiring along bottom plate at exterior wall *

Caulk/seal/foam all electrical wires penetrating into vented crawl spaces or into garages

Caulk/seal/foam all electrical wires penetrating into exterior wall

Figure 8.3
Sealing Electrical Wires

- Run low voltage wires in plastic conduit to allow for future upgrade or service
* Some codes require wires to be held up from bottom plates 6" to 8" to protect wires from future drilling of holes through plates

Avoid placing recessed lights in insulated ceilings unless they are specifically designed to be airtight. Install IC-rated fixtures that have passed the ASTM E-283 test for air leakage.

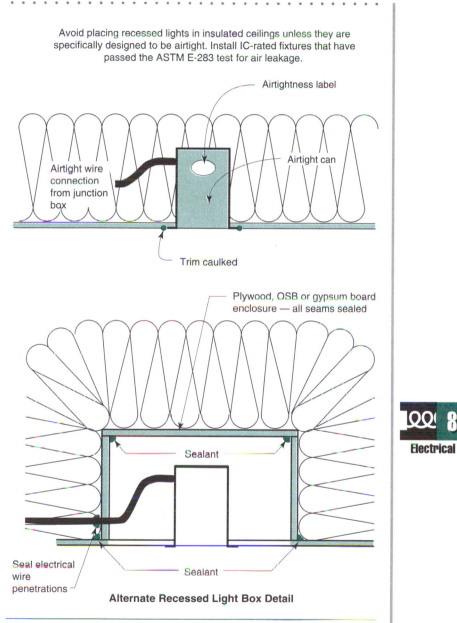

Airtightness label

Airtight can

Airtight wire connection from junction box

Trim caulked

Plywood, OSB or gypsum board enclosure — all seams sealed

Sealant

Seal electrical wire penetrations

Sealant

Alternate Recessed Light Box Detail

8

Electrical

Figure 8.4
Airtight Recessed Light Box

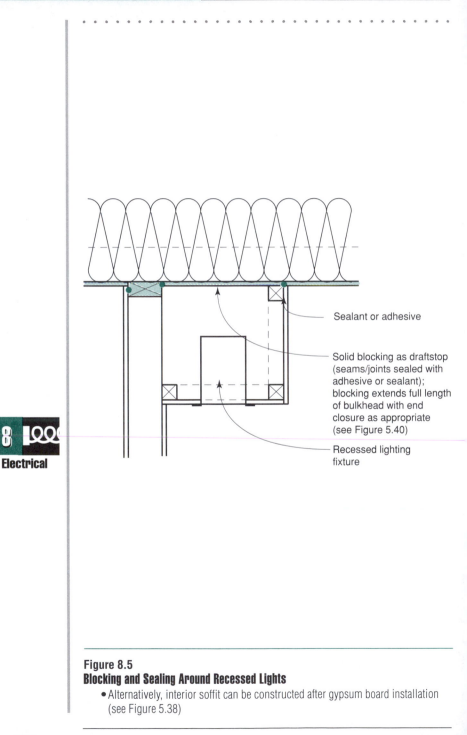

Sealant or adhesive

Solid blocking as draftstop (seams/joints sealed with adhesive or sealant); blocking extends full length of bulkhead with end closure as appropriate (see Figure 5.40)

Recessed lighting fixture

8

Electrical

Figure 8.5
Blocking and Sealing Around Recessed Lights
- Alternatively, interior soffit can be constructed after gypsum board installation (see Figure 5.38)

Insulation

Cavity insulation combined with insulating sheathings are common in residential wall construction. Cavity insulations are typically fiberglass batt, damp spray cellulose, dry spray cellulose, fiberglass, rock or slag wool supported by netting or reinforced polyethylene and spray foam. Insulating sheathings are typically extruded and expanded polystyrenes, foil- and fiber-faced isocyanurates, and rigid fiberglass. Roof insulations are typically blown fiberglass, blown cellulose, fiberglass batt, and spray foam. All have to:

- keep the heat in during the winter
- keep the heat out during the summer

Concerns

Fiberglass batt, damp spray cellulose, dry spray cellulose, blown fiberglass, rock or slag wool and blown cellulose cavity and roof insulations are not air flow retarders. They should be used in conjunction with air flow retarders. Spray foams and rigid insulations (when their joints are sealed) are air flow retarders in their own right.

Insulation

Just so that we are all clear, blowing a cathedral ceiling "solid" with cellulose will not eliminate the need for an interior air flow retarder or the need for roof ventilation if the temperature of the first condensing surface (underside of roof deck) is not controlled by the use of rigid insulation. Packing fiberglass around window frames or around plumbing stacks to reduce air flow, while better than nothing at all, does not effectively stop air flow. Foam sealants or caulks provide effective air sealing in these areas.

Let's also point out that properly treated cellulose is not, and we repeat is not, more of a fire hazard than fiberglass insulation. Furthermore, regarding health related risks associated with different insulation types, loose fibers and particulates of many sizes and *many* materials (fiberglass, rock or slag wool, cellulose) have been known to irritate many people. As long as proper precautions are taken during installation and proper containment and air sealing of these insulations is made, the health risk is negligeable compared to the benefits provided by the energy savings.

To be perfectly clear: insulation is good. All insulation is good. More insulation is better. All insulation is environmentally friendly, even the rigid foams because of the shear quantity of energy (barrels of oil not burned, pounds of carbon not dumped into the atmosphere) they save over their useful service lives. The amount of energy used to make all insulation (the embodied energy, even in the rigid foams) is trivial compared to the energy they save when used in buildings that last at least one mortgage period (don't rot and fall down). So the key to environmentally friendly construction is durable building envelopes that are extremely well insulated.

Fiberglass

In wall cavities, fiberglass batt insulation should be cut to fit and carefully installed to completely fill the cavity. Batts should not be cut short or cut long and forced/compressed into small areas. Batts should be fluffed to full thickness and split around plumbing and wiring (Figure 9.2). Even better, move the wires so that insulation does not have to be split. It is preferable to face staple fiberglass batts, with integral vapor retarders as inset stapling can affect performance (Figure 9.1). Yes, face stapled batts can interfere with drywall installation. If you now need a vapor diffusion retarder because you are not using faced batts use paint. The use of higher density batts will typically improve the quality of installation.

9

Insulation

Cellulose

Damp spray cellulose insulation should only be used in wall assemblies that are able to dry towards the interior or toward the exterior. If a vapor impermeable sheathing is installed on the exterior of a wall assembly, a vapor diffusion retarder should not be installed on the interior. Dry spray cellulose can be used in wall cavities with netting and can be used with any type of sheathing. Cellulose is not a vapor diffusion retarder. That means if you need a vapor diffusion retarder on the interior of a building assembly you should use polyethylene or low perm (vapor retarder) paint. Keep in mind that polyethylene on the inside of building assemblies in mixed-humid and mixed-dry climates is not generally a good idea (see Appendix IV) and a definite no in hot-humid and hot-dry climates.

Roofs

In all truss and roof assemblies, baffles should be installed at roof perimeters to prevent the wind-washing of thermal insulation and to prevent insulation from blocking soffit vents in vented roof assemblies (Figure 9.3 and 9.4).

Ventilation can be used to remove moisture from roof assemblies and to control ice damming. Attic ventilation is most effective when half the vent area is near the ridge and half is near the eave. Typical practice requires 1-square-foot of net free vent area for every 300-square-feet of ceiling area (Figure 9.5).

In mixed-humid, mixed-dry, hot-humid and hot-dry climates it is not recommended that a vapor diffusion retarder be installed in vented roof assemblies. During hot, humid weather, hot, humid ventilation air is brought into vented roof assemblies. Under such conditions roof assembly vapor diffusion retarders are typically on the "wrong" side of the assembly (i.e. towards the interior). It is better to not have a ceiling vapor diffusion retarder to permit drying to the interior. Under heating conditions, roof ventilation will flush moisture migrating from the interior out of vented roof assemblies.

Crawl Spaces

Crawl spaces should not be vented. Crawl spaces should be constructed as mini basements. They should be part of the conditioned space of the house. That means that they should have a supply HVAC system duct. A return HVAC system duct or grille is not typically recommended — the "leakiness" of the floor assembly will provide the return air path. However, a return duct, dampered to prevent crawl space depressurization (via excessive return air flow) is acceptable. Alternatively, a floor grille acting as a "transfer" grille can be installed. It should be noted installation of a "transfer" grille between the crawl space and the house may be in conflict with some codes unless crawl space services are also fire rated. The crawl space may be considered a plenum with such an approach. Accidental "active" depressurization of the conditioned crawl spaces via a return duct or return duct leakage that creates recirculation into the house should be avoided.

Insulation

Spray Foam

Spray foam insulations provide excellent air sealing characteristics and can be used to provide air flow retarder continuity at difficult details such as across band braces. Use of low density foams result in flexible installations, forgiving of movement. Higher density foams are more abuse-resistant but are not as tolerant of movement.

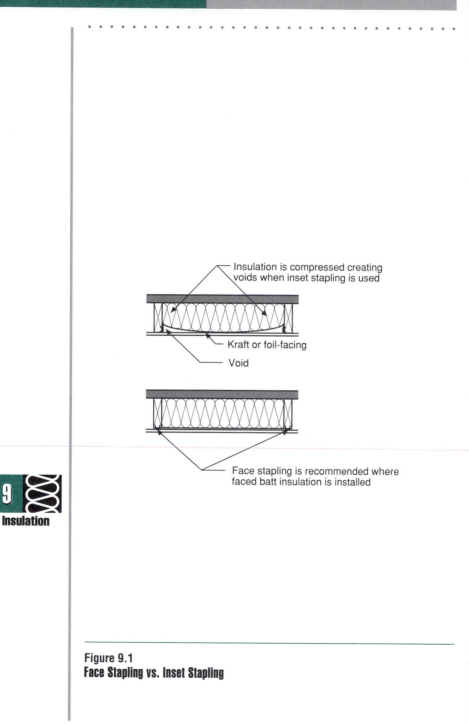

Insulation is compressed creating voids when inset stapling is used

Kraft or foil-facing

Void

Face stapling is recommended where faced batt insulation is installed

9

Insulation

Figure 9.1
Face Stapling vs. Inset Stapling

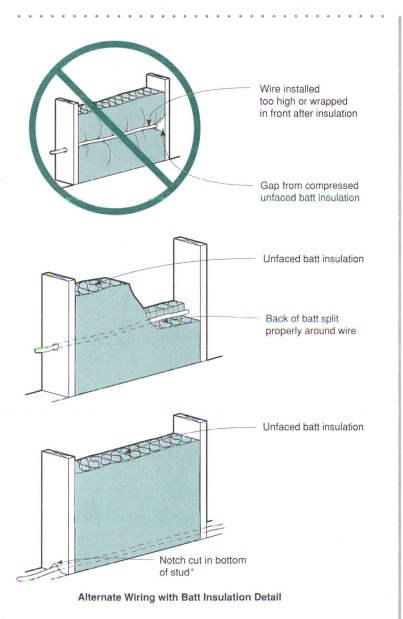

Wire installed
too high or wrapped
in front after insulation

Gap from compressed
unfaced batt insulation

Unfaced batt insulation

Back of batt split
properly around wire

Unfaced batt insulation

Notch cut in bottom
of stud*

Alternate Wiring with Batt Insulation Detail

9

Insulation

Figure 9.2
Installing Batt Insulation in Cavity with Electrical Wiring
  * Some codes require wires to be held up from bottom plate 6" to 8" to protect
  wires from future drilling of holes through plates

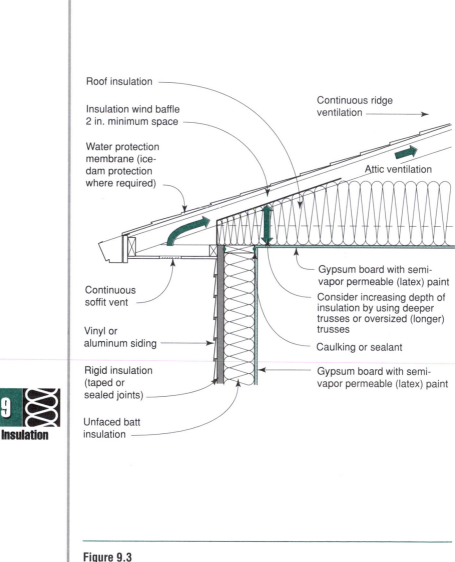

Roof insulation

Insulation wind baffle 2 in. minimum space

Water protection membrane (ice-dam protection where required)

Continuous ridge ventilation

Attic ventilation

Continuous soffit vent

Vinyl or aluminum siding

Rigid insulation (taped or sealed joints)

Unfaced batt insulation

Gypsum board with semi-vapor permeable (latex) paint

Consider increasing depth of insulation by using deeper trusses or oversized (longer) trusses

Caulking or sealant

Gypsum board with semi-vapor permeable (latex) paint

Figure 9.3
Baffle Installation

- Roof insulation thermal resistance (depth) at truss heel (roof perimeter) should be equal or greater to thermal resistance of exterior wall

9

Insulation

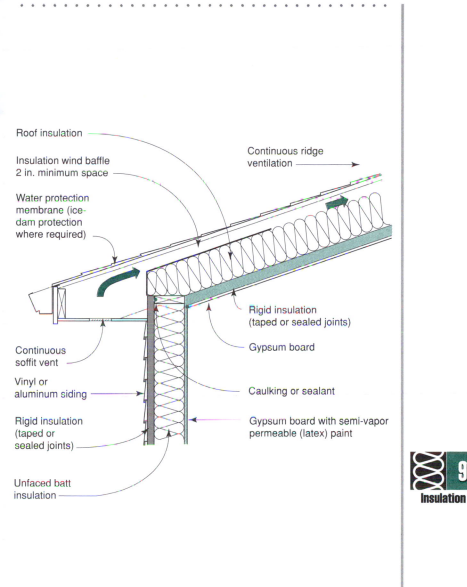

Roof insulation

Insulation wind baffle
2 in. minimum space

Water protection
membrane (ice-
dam protection
where required)

Continuous ridge
ventilation

Rigid insulation
(taped or sealed joints)

Gypsum board

Continuous
soffit vent

Vinyl or
aluminum siding

Caulking or sealant

Rigid insulation
(taped or
sealed joints)

Gypsum board with semi-vapor
permeable (latex) paint

Unfaced batt
insulation

9

Insulation

Figure 9.4
Baffle Installation in a Cathedral Ceiling
- Under hot, humid conditions, ventilation air brought into the roof assembly
  from the exterior will not condense on the upper surface of the rigid insulation
  (if the rigid insulation is a vapor diffusion retarder) if the R-value of the
  insulation is sufficient; R-7.5 or higher is recommended for mixed climates.

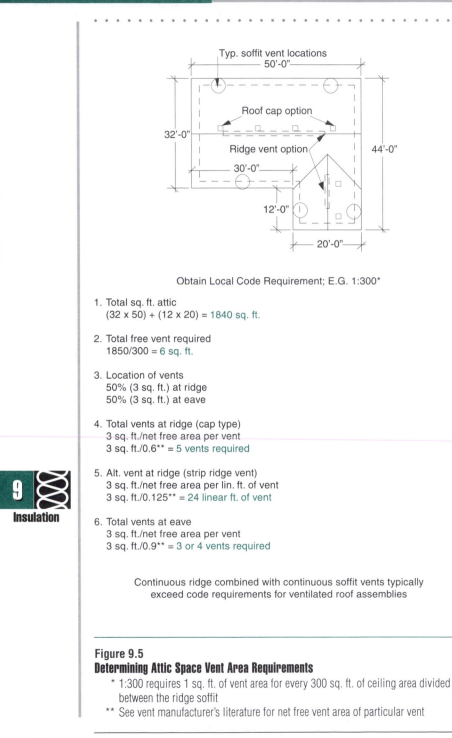

Obtain Local Code Requirement; E.G. 1:300*

1. Total sq. ft. attic
   (32 x 50) + (12 x 20) = 1840 sq. ft.

2. Total free vent required
   1850/300 = 6 sq. ft.

3. Location of vents
   50% (3 sq. ft.) at ridge
   50% (3 sq. ft.) at eave

4. Total vents at ridge (cap type)
   3 sq. ft./net free area per vent
   3 sq. ft./0.6** = 5 vents required

5. Alt. vent at ridge (strip ridge vent)
   3 sq. ft./net free area per lin. ft. of vent
   3 sq. ft./0.125** = 24 linear ft. of vent

6. Total vents at eave
   3 sq. ft./net free area per vent
   3 sq. ft./0.9** = 3 or 4 vents required

   Continuous ridge combined with continuous soffit vents typically
   exceed code requirements for ventilated roof assemblies

9

Insulation

Figure 9.5
Determining Attic Space Vent Area Requirements
 * 1:300 requires 1 sq. ft. of vent area for every 300 sq. ft. of ceiling area divided
   between the ridge soffit
** See vent manufacturer's literature for net free vent area of particular vent

Drywall

Drywall (gypsum board) has to:

- provide rigidity
- provide aesthetics
- provide fire protection
- not leak air

Concerns

Wood moves. Drywall does not move. Interesting problem. The more you attach drywall to wood, the more cracks you have. Easy, attach the drywall to less wood, and, in a manner that allows the wood to move.

Nail pops happen because as wood dries, it shrinks. Nails do not shrink. Actually, nails do not pop. The wood shrinks away from the back face of the drywall as it dries. How about getting dry wood? Sure. Better to use shorter nails. Even better, use glue. With glue, as the wood shrinks, it pulls the drywall inwards with it. But, you can't only use glue, you've got to use something until the glue begins to work. Now, shorter nails don't hold very well, and we don't want to use more of them, so use shorter screws and glue.

Wood is weird. When it shrinks, it shrinks differently along the grain than perpendicular to the grain. It shrinks much more at right angles to the grain, than along the grain. Studs don't get shorter, but they get thinner in thickness and in width (Figure 10.1 and 10.2). When we attach drywall, we need to keep this in mind especially when we box in built-up beams made out of 2x10s and 2x12s. What's nice about engineered wood, is it doesn't shrink. Drywall likes engineered wood, especially above windows as header material. Drywall doesn't like big pieces of real wood.

Drywall

Stairwells provide an interesting problem. With real wood floor joists (2x10s), you get more shrinkage in the $9^1/_4$ inches of floor framing (remember this is at right angles to the grain of the floor joists) than in the 8 feet of wall framing (along the grain of the studs) above and below. Old timers used to balloon frame two story spaces for this reason, or provide control joints between the floors in the plaster or drywall. Better to use floor trusses or engineered wood joists, they don't shrink.

Truss Uplift

What can we say about truss uplift? You can't prevent it. Truss uplift occurs because of moisture content differences between the upper and lower chords of wood trusses. Moisture content differences are inevitable if one member is cold and the other member is warm. If you insulate a wood roof truss, the lower chords will be warm and the upper ones will be cold. Remember, truss uplift is not truss uplift if the owner can't see it. Let the trusses move. Floating corners for drywall attachment is the way to go. The truss moves, the drywall bends, no crack, end of story (Figures 10.4 and 10.5). This is also the same principle to use at corners of exterior and interior walls. Why use three stud corners? If we attach the drywall to the wood on both sides of the corner, when the wood shrinks, the drywall cracks. Two stud corners are better. Don't attach the drywall (Figure 10.3). Let the wood move. If you are going to use a three stud corner, at least don't attach the drywall to one of the sides, just support the drywall until it is taped. Let the tape hold the corner together.

Air Flow Retarders

Drywall

One of the really nice things about drywall is that air doesn't leak through it. It leaks around it, but not through it. Tape the joints of drywall together, glue it to top and bottom plates and around window openings and presto - you have an air flow retarder (Figures 10.6 through 10.8). Now paint it and shazzam - you have a vapor diffusion retarder. Add draftstops and firestops out of rigid material to the framing package and we have an airtight building.

Now drywall is not the only air flow retarder. Polyethylene installed in a continuous fashion (seams taped or sealed) works as an air flow retarder. So does spray foam. If you are using polyethylene, it is very important that the installation of the drywall does not cause rips and tears in the poly. Keep in mind, however, polyethylene on the inside of building assemblies in mixed-humid climates is not generally a good idea (see Appendix IV).

Ceramic Tile Tub and Shower Enclosures

Treated gypsum board ("green board") is typically used as a base for ceramic tile enclosures around bathtubs and showers. This type of gypsum board (or any gypsum board) should not be installed over polyethylene in an exterior insulated wall under tile. Once again, it should be noted that using polyethylene on the inside of a wall assembly in a mixed-humid climate is generally not a good idea. Moisture becomes trapped between the tile and the poly. The gypsum board goes to mush (green board included, it just takes longer). Moisture gets in because grout joints are permeable to moisture. Using cement boards or cement and wire lath in place of gypsum board is always a good idea.

Winter Construction

Drywall under insulated ceilings should be $^5/_8$-inch thick. Interior and exterior walls should be framed on 24-inch centers and standard $^1/_2$-inch drywall is fine. Don't tape under insulated ceilings in the winter if you haven't insulated. Big mistake. Try it and you will learn all about condensation and how expensive it is to put up new drywall. Yes, but I really like blown insulation and I can't blow until my drywall is up. Well, you can board first, then blow and then tape. Or, you can use a batt and blow strategy. Batt the ceiling with a little insulation when you do the walls, and then blow the loose stuff on top later.

Winter construction is always fun. Propane heaters release lots and lots of moisture. Interesting problem. You want to dry out your building while you are humidifying it. Okay, so we open some windows. Let's now heat a building with open windows. Propane heaters also release lots and lots of carbon dioxide. Joint compound does not like carbon

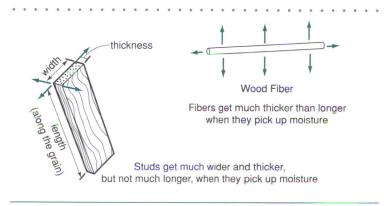

thickness — width — length (along the grain)

Wood Fiber

Fibers get much thicker than longer when they pick up moisture

Studs get much wider and thicker, but not much longer, when they pick up moisture

Figure 10.1
Wood Shrinkage

Drywall

dioxide. Bad things happen with lots of carbon dioxide, moisture and joint compound; carbonation. Carbonation is bad. Better to hook your heating system up. If you have gas heat, make sure you have a chimney. The alternative is to install temporary heat, properly vented. You still need air change to flush out the moisture, but not as much as before. Moisture is bad, carbon dioxide is bad, heat and ventilation are good. Use a setting type compound with humid or cool conditions. These compounds can be selected for a range of specific properties.

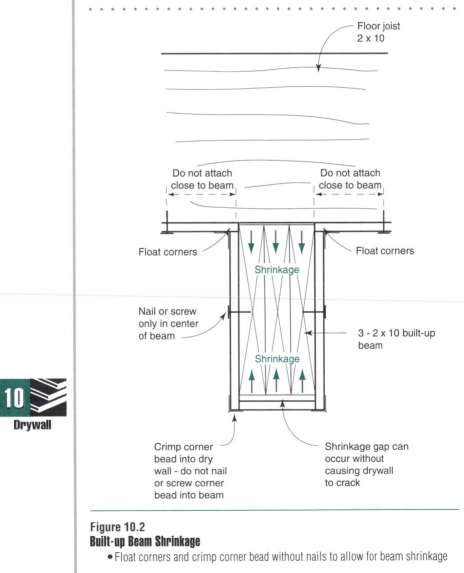

Figure 10.2
Built-up Beam Shrinkage
  • Float corners and crimp corner bead without nails to allow for beam shrinkage

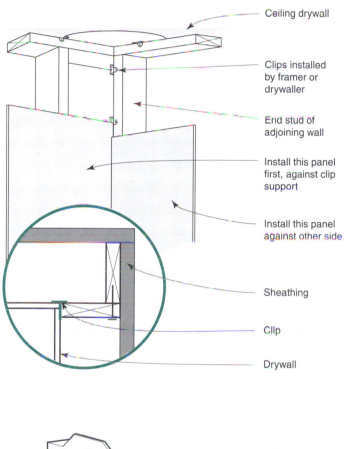

Ceiling drywall

Clips installed
by framer or
drywaller

End stud of
adjoining wall

Install this panel
first, against clip
support

Install this panel
against other side

Sheathing

Clip

Drywall

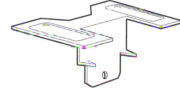

10

Drywall

Figure 10.3
Typical Clip Support for Gypsum Board

• Use of clip support for gypsum board results in floating corners and
significantly less drywall cracking

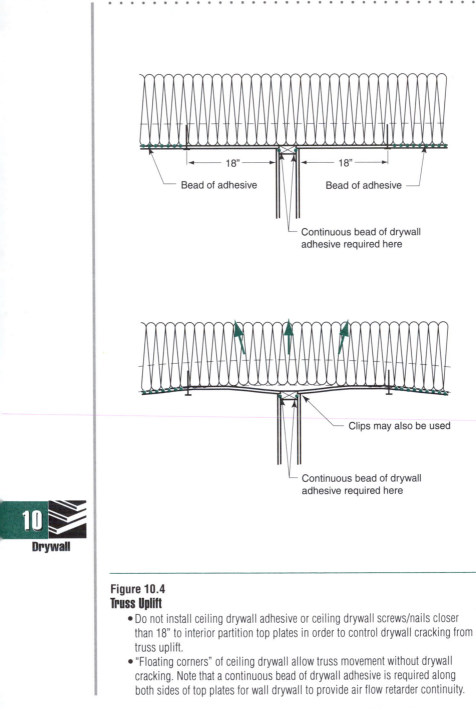

Figure 10.4
Truss Uplift

- Do not install ceiling drywall adhesive or ceiling drywall screws/nails closer than 18" to interior partition top plates in order to control drywall cracking from truss uplift.
- "Floating corners" of ceiling drywall allow truss movement without drywall cracking. Note that a continuous bead of drywall adhesive is required along both sides of top plates for wall drywall to provide air flow retarder continuity.

10

Drywall

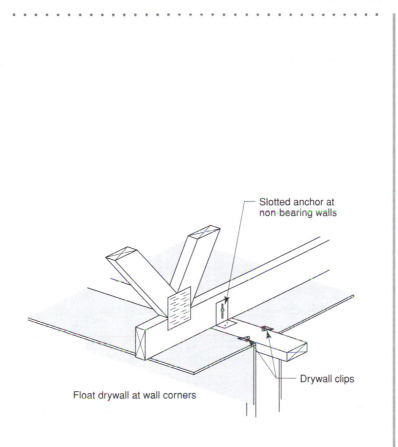

Slotted anchor at
non-bearing walls

Drywall clips

Float drywall at wall corners

Drywall

Figure 10.5
Drywall Clips and Slotted Anchor on Non-Bearing Wall

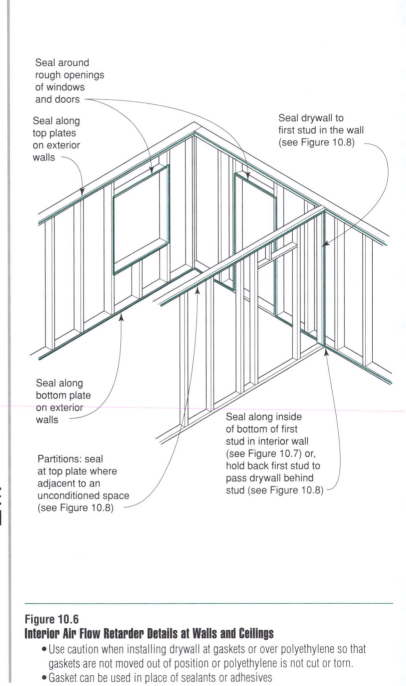

Seal around rough openings of windows and doors

Seal along top plates on exterior walls

Seal drywall to first stud in the wall (see Figure 10.8)

Seal along bottom plate on exterior walls

Seal along inside of bottom of first stud in interior wall (see Figure 10.7) or, hold back first stud to pass drywall behind stud (see Figure 10.8)

Partitions: seal at top plate where adjacent to an unconditioned space (see Figure 10.8)

10

Drywall

Figure 10.6
Interior Air Flow Retarder Details at Walls and Ceilings

- Use caution when installing drywall at gaskets or over polyethylene so that gaskets are not moved out of position or polyethylene is not cut or torn.
- Gasket can be used in place of sealants or adhesives

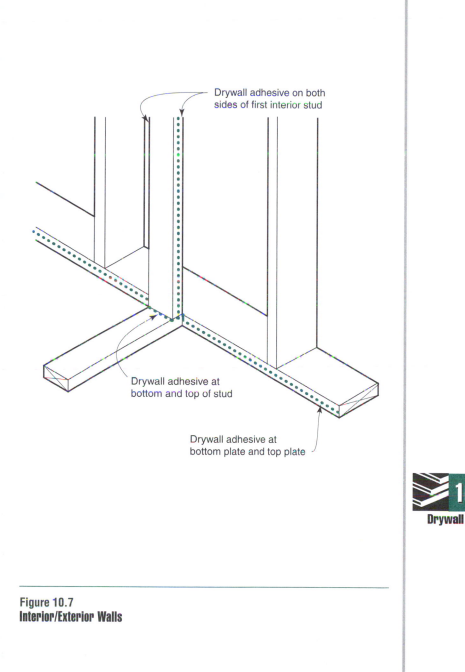

Drywall adhesive on both
sides of first interior stud

Drywall adhesive at
bottom and top of stud

Drywall adhesive at
bottom plate and top plate

10
Drywall

Figure 10.7
Interior/Exterior Walls

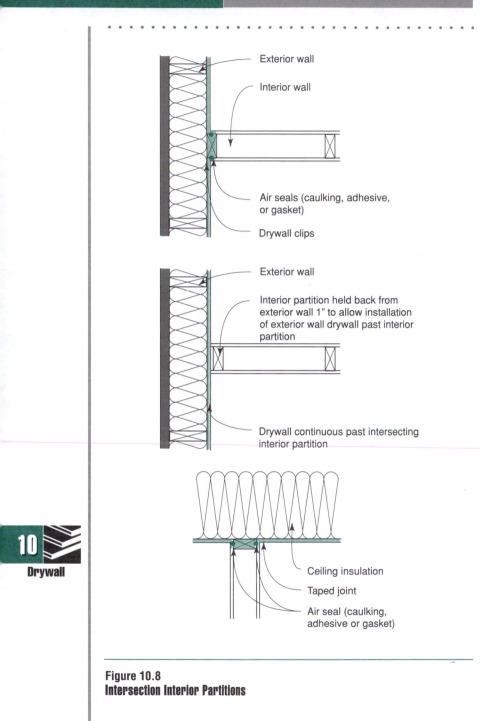

Exterior wall

Interior wall

Air seals (caulking, adhesive, or gasket)

Drywall clips

Exterior wall

Interior partition held back from exterior wall 1" to allow installation of exterior wall drywall past interior partition

Drywall continuous past intersecting interior partition

Ceiling insulation

Taped joint

Air seal (caulking, adhesive or gasket)

10

Drywall

Figure 10.8
Intersection Interior Partitions

Painting

Paint has to:

- keep rain out of wood

- breathe when it's on the outside

- protect wood from getting a sunburn

- look nice

Concerns

Exposure to sunlight (ultraviolet radiation) and moisture are the major factors affecting the durability of paint coatings and the durability of wood. Although, independently each factor can lead to deterioration, the effect of the combination of both factors is much more severe than either factor separately. Ultraviolet radiation and moisture can each lead to the breakdown of the resin in painted surfaces which binds (holds) the pigment to the surface. When the resin breaks down, pigment is lost (washed away from the surface) and fading occurs. In some instances, rubbing the surface with a cloth or a hand will remove a white powder from the paint surface (chalking). Ultraviolet radiation and moisture have a similar effect on wood. Wood breaks down under exposure to ultraviolet radiation and wood surface erosion is increased with exposure to ultraviolet radiation coupled with rain.

The ideal coating system for wood is a system which is hydrophobic (sheds water), vapor permeable (breathes), resistant to ultraviolet light (sunlight) and has good adhesion (sticks to wood) and cohesion (stretches) properties.

Acrylic latex top coats coupled with premium latex primers are recommended when they are applied over stable substrates (dry, dimensionally stable and able to hold paint) as they are more vapor permeable

Painting

than other paint finishes while providing similar hydrophobic, UV resistance, adhesion and cohesion properties. Two coats of all-acrylic latex paint over a premium latex primer are recommended.

The optimum thickness for the total dry paint coat (primer and two top-coats) is 3.5 to 5 mils.

Oil-based prime coats coupled with latex top coats do not provide as permeable a system as a latex prime coat based system. However, oil-based prime coats provide superior adhesion and stain blocking characteristics for difficult substrates. For woods with water-soluble extractives, such as redwood and cedar, oil-based prime coats are recommended. Do not use latex-based prime coats with these type of substrates.

Exposure of unprotected wood to sunlight can adversely affect the adhesion of paint to wood within as little as 3-to-4 week exposures. Wood surfaces should be painted as soon as possible, weather permitting. All exterior wood (except decking materials) should be back primed or prime coated on all six surfaces. Ideally, wood should be pre-primed on all surfaces prior to arrival at the job site. Field cut edges should be sealed with primer during installation. Top coats should be applied within 2 weeks of field exposure of the prime coat. The sooner, the better. Some prime coats weather by forming a soap-like film that can interfere with adhesion of top coats. Washing of aged prime coats (exposed to sunlight) is recommended prior to top coat application. Re-priming may be necessary if prime coats have excessively weathered. Ideally, the temperature should not drop below 50 degrees for at least 24 hours after paint application. Winter, late fall or early spring topcoat application is not recommended.

Pre-primed material should be utilized during winter construction and not top-coated (finished) until weather permits.

A paint coating's resistance to ultraviolet radiation and moisture is dependent on the ratio of resin to pigment in the paint. The more resin available to completely coat a pigment particle, the more forcefully the particle is bound to a surface. Premium paints have a high ratio of resin to pigment. A low cost paint typically has a high pigment content relative to resin content as pigment is less expensive than resin. Although a high pigment content paint has an excellent "hiding" ability, high pigment content paints with low resin contents are unable to resist exposure to sunlight and moisture. Gloss paints have more resin than semi-gloss paints, and semi-gloss paints have more resin than flat paints. Gloss paints have the most resistance to ultraviolet radiation and moisture, flat paints have the least.

11

Painting

Stains are not as hydrophobic or resistant to ultraviolet light as paints but are more vapor permeable. Since stains break down more rapidly due to ultraviolet light than do paints, re-coating more frequently with stains will be likely. Solid body stains are thin paints and should not be used. Do not use solid body stains.

Deck materials should never be painted as even vapor permeable paint coatings serve to inhibit drying of absorbed moisture beyond acceptable levels for such a hostile moisture environment (horizontal, exposed to rain and sun). However, deck materials can be coated with penetrating water repellents or stains. Both of these serve to reduce water absorption without reducing drying ability. All surfaces should be treated including the back. Untreated deck materials will deteriorate from both water absorption and exposure to ultraviolet light. Even preservative treated deck materials deteriorate from exposure to ultraviolet light. Stains provide satisfactory protection against UV exposure (sunlight). Straight water repellents do not. Stains act as a type of "suntan lotion" for the wood. Like most typical sun tan lotions, stains must be regularly reapplied. Superior performance may be achieved when stains are applied over preservative treated decking materials.

Interior Surfaces

Paint coatings installed on interior surfaces can be either permeable or impermeable depending on the design of the wall assembly. On wall and basement assemblies which are designed to dry to the interior, only vapor permeable paint systems (latex paint) should be used.

On wall and roof assemblies which require a surface applied interior vapor diffusion retarder, a vapor impermeable paint system should be used. These coatings are available in both latex-based and oil-based systems.

Paints with low or no emissions of volatile organic compounds (VOC's) should be selected for interior application to reduce concentrations of interior contaminants.

11

Painting

11

Painting

Appendix I

Geographic Data

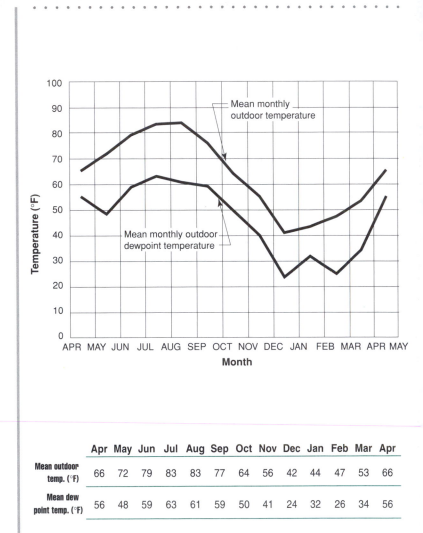

	Apr	May	Jun	Jul	Aug	Sep	Oct	Nov	Dec	Jan	Feb	Mar	Apr
Mean outdoor temp. (°F)	66	72	79	83	83	77	64	56	42	44	47	53	66
Mean dew point temp. (°F)	56	48	59	63	61	59	50	41	24	32	26	34	56

Figure I.1
Abilene, Texas
- Average annual precipitation 19.2 inches
- 2,624 heating degree days
- Winter design temperature 20°F
- Summer design temperature 99°F dry bulb, 71°F wet bulb
- Average deep ground temperature 63°F
- Latitude 32.43°

12 I

Appendices

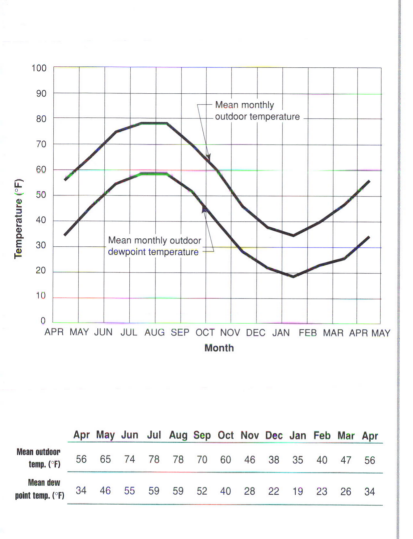

	Apr	May	Jun	Jul	Aug	Sep	Oct	Nov	Dec	Jan	Feb	Mar	Apr
Mean outdoor temp. (°F)	56	65	74	78	78	70	60	46	38	35	40	47	56
Mean dew point temp. (°F)	34	46	55	59	59	52	40	28	22	19	23	26	34

Figure I.2
Amarillo, Texas
- Average annual precipitation 20 inches
- 4,258 heating degree days; 1,354 cooling degree days
- Winter design temperature 11°F
- Summer design temperature 95°F dry bulb, 67°F wet bulb
- Average deep ground temperature 57°F
- Latitude 35.13°

I 12

Appendices

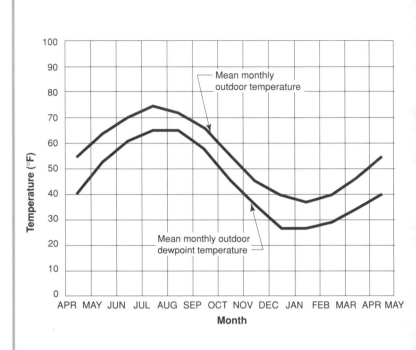

	Apr	May	Jun	Jul	Aug	Sep	Oct	Nov	Dec	Jan	Feb	Mar	Apr
Mean outdoor temp. (°F)	55	63	70	74	72	66	56	46	40	37	40	47	55
Mean dew point temp. (°F)	40	52	61	65	65	58	46	36	27	27	29	34	40

Figure I.3
Asheville, North Carolina
- Average annual precipitation 47.6 inches
- 4,308 heating degree days; 787 cooling degrees days
- Winter design temperature 14°F
- Summer design temperature 85°F dry bulb, 71°F wet bulb
- Average deep ground temperature 55°F
- Latitude 35.26°

Appendices

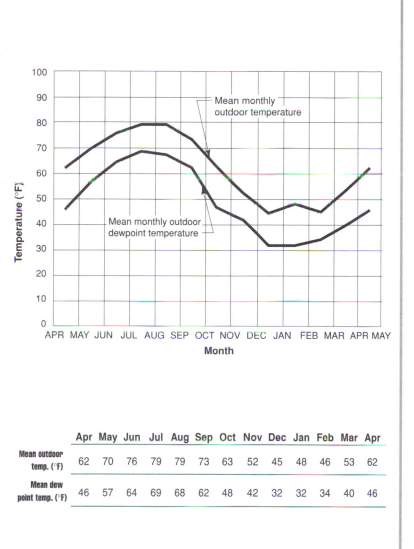

	Apr	May	Jun	Jul	Aug	Sep	Oct	Nov	Dec	Jan	Feb	Mar	Apr
Mean outdoor temp. (°F)	62	70	76	79	79	73	63	52	45	48	46	53	62
Mean dew point temp. (°F)	46	57	64	69	68	62	48	42	32	32	34	40	46

Figure I.4
Atlanta, Georgia
- Average annual precipitation 51 inches
- 2,991 heating degree days; 1,667 cooling degrees days
- Winter design temperature 22°F
- Summer design temperature 92°F dry bulb, 74°F wet bulb
- Average deep ground temperature 55°F
- Latitude 33.38°

I **12**

Appendices

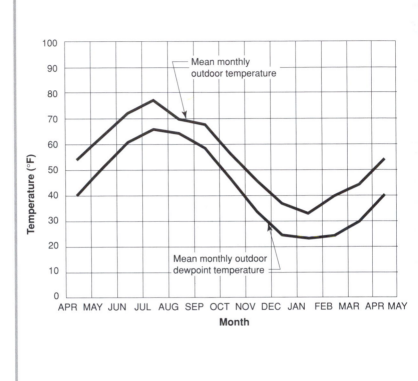

	Apr	May	Jun	Jul	Aug	Sep	Oct	Nov	Dec	Jan	Feb	Mar	Apr
Mean outdoor temp. (°F)	54	63	72	77	75	68	57	46	37	33	35	43	54
Mean dew point temp. (°F)	40	51	61	67	65	58	47	34	25	23	24	30	40

Figure I.5
Baltimore, Maryland
- Average annual precipitation 40.8 inches
- 4,654 heating degree days; 1,137 cooling degree days
- Winter design temperature 13°F
- Summer design temperature 91°F dry bulb, 75°F wet bulb
- Average deep ground temperature 55°F
- Latitude 39.10°

12 **I**

Appendices

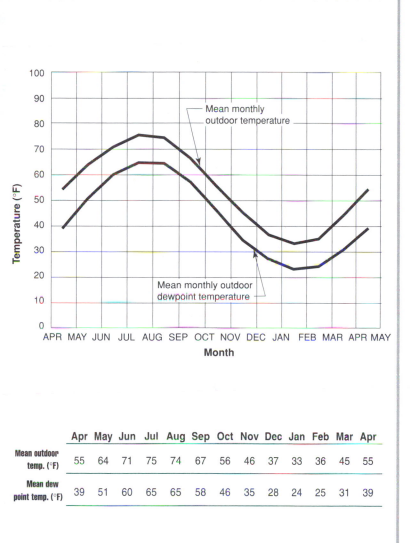

	Apr	May	Jun	Jul	Aug	Sep	Oct	Nov	Dec	Jan	Feb	Mar	Apr
Mean outdoor temp. (°F)	55	64	71	75	74	67	56	46	37	33	36	45	55
Mean dew point temp. (°F)	39	51	60	65	65	58	46	35	28	24	25	31	39

Figure I.6
Charleston, West Virginia
- Average annual precipitation 42.5 inches
- 4,476 heating degree days; 1,031 cooling degrees days
- Winter design temperature 11°F
- Summer design temperature 90°F dry bulb, 73°F wet bulb
- Average deep ground temperature 55°F
- Latitude 38.22°

Appendices

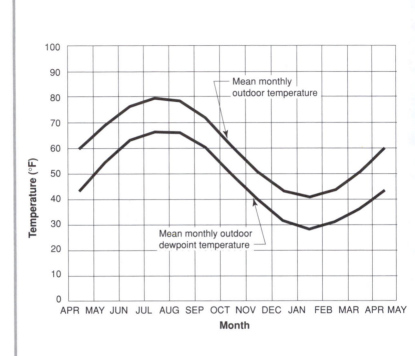

	Apr	May	Jun	Jul	Aug	Sep	Oct	Nov	Dec	Jan	Feb	Mar	Apr
Mean outdoor temp. (°F)	60	69	76	79	78	72	61	51	43	41	44	51	60
Mean dew point temp. (°F)	43	55	63	67	67	61	50	40	32	28	31	36	43

Figure I.7
Charlotte, North Carolina
- Average annual precipitation 43.1 inches
- 3,341 heating degree days; 1,582 cooling degree days
- Winter design temperature 22°F
- Summer design temperature 93°F dry bulb, 74°F wet bulb
- Average deep ground temperature 61°F
- Latitude 35.13°

Appendices

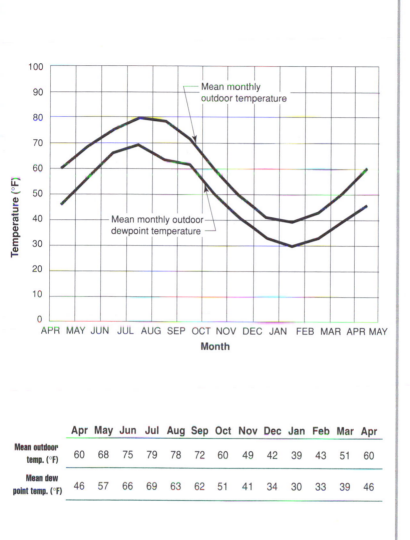

	Apr	May	Jun	Jul	Aug	Sep	Oct	Nov	Dec	Jan	Feb	Mar	Apr
Mean outdoor temp. (°F)	60	68	75	79	78	72	60	49	42	39	43	51	60
Mean dew point temp. (°F)	46	57	66	69	63	62	51	41	34	30	33	39	46

Figure I.8
Chattanooga, Tennessee
- Average annual precipitation 53.5 inches
- 3,587 heating degree days; 1,544 cooling degree days
- Winter design temperature 18°F
- Summer design temperature 93°F dry bulb, 74°F wet bulb
- Average deep ground temperature 60°F
- Latitude 35.02°

Appendices

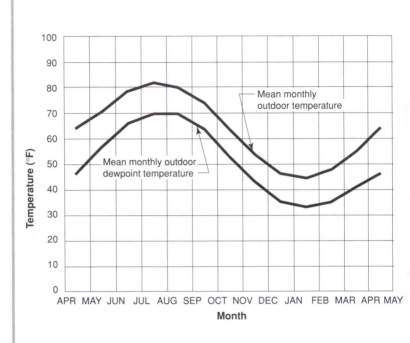

	Apr	May	Jun	Jul	Aug	Sep	Oct	Nov	Dec	Jan	Feb	Mar	Apr
Mean outdoor temp. (°F)	63	71	78	82	80	74	64	54	47	45	48	55	63
Mean dew point temp. (°F)	47	57	66	70	70	64	53	44	36	33	36	41	47

Figure I.9
Columbia, South Carolina

- Average annual precipitation 50 inches
- 2,649 heating degree days; 1,966 cooling degree days
- Winter design temperature 24°F
- Summer design temperature 95°F dry bulb, 75°F wet bulb
- Average deep ground temperature 63°F
- Latitude 33.56°

Appendices

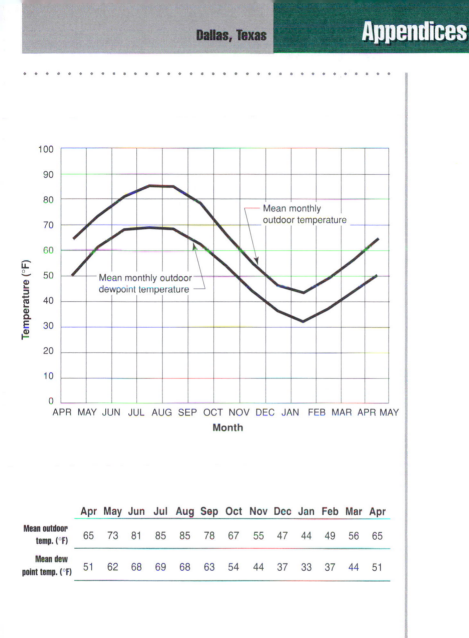

	Apr	May	Jun	Jul	Aug	Sep	Oct	Nov	Dec	Jan	Feb	Mar	Apr
Mean outdoor temp. (°F)	65	73	81	85	85	78	67	55	47	44	49	56	65
Mean dew point temp. (°F)	51	62	68	69	68	63	54	44	37	33	37	44	51

Figure I.10
Dallas, Texas

- Average annual precipitation 33.7 inches
- 2,407 heating degree days; 2,603 cooling degree days
- Winter design temperature 22°F
- Summer design temperature 100°F dry bulb, 75°F wet bulb
- Average deep ground temperature 66°F
- Latitude 32.53°

Appendices

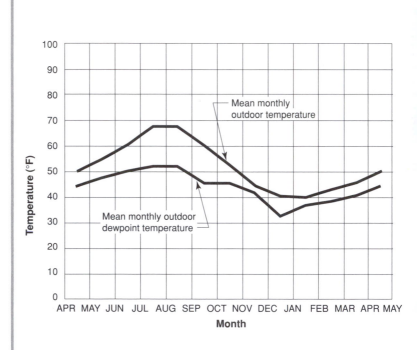

	Apr	May	Jun	Jul	Aug	Sep	Oct	Nov	Dec	Jan	Feb	Mar	Apr
Mean outdoor temp. (°F)	50	55	61	67	67	61	53	45	41	40	43	46	50
Mean dew point temp. (°F)	44	47	50	52	52	46	46	42	33	37	38	41	44

Figure I.11
Eugene, Oregon

- Average annual precipitation 49.4 inches
- 4,546 heating degree days; 300 cooling degree days
- Winter design temperature 22°F
- Summer design temperature 89°F dry bulb, 66°F wet bulb
- Average deep ground temperature 52°F
- Latitude 44.08°

12 **I**

Appendices

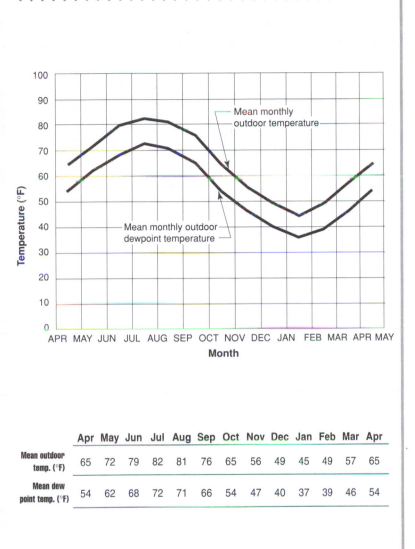

	Apr	May	Jun	Jul	Aug	Sep	Oct	Nov	Dec	Jan	Feb	Mar	Apr
Mean outdoor temp. (°F)	65	72	79	82	81	76	65	56	49	45	49	57	65
Mean dew point temp. (°F)	54	62	68	72	71	66	54	47	40	37	39	46	54

Figure I.12
Jackson, Mississippi
- Average annual precipitation 55.4 inches
- 2,467 heating degree days; 2,215 cooling degree days
- Winter design temperature 25°F
- Summer design temperature 95°F dry bulb, 76°F wet bulb
- Average deep ground temperature 64°F
- Latitude 32.19°

Appendices

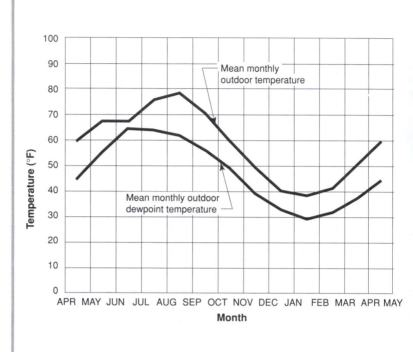

	Apr	May	Jun	Jul	Aug	Sep	Oct	Nov	Dec	Jan	Feb	Mar	Apr
Mean outdoor temp. (°F)	59	67	67	75	78	71	60	49	41	38	42	50	59
Mean dew point temp. (°F)	45	56	64	63	62	56	49	39	33	29	32	37	45

Figure I.13
Knoxville, Tennessee

- Average annual precipitation 47.1 inches
- 3,937 heating degree days; 1,266 cooling degree days
- Winter design temperature 19°F
- Summer design temperature 92°F dry bulb, 73°F wet bulb
- Average deep ground temperature 59°F
- Latitude 35.49°

Appendices

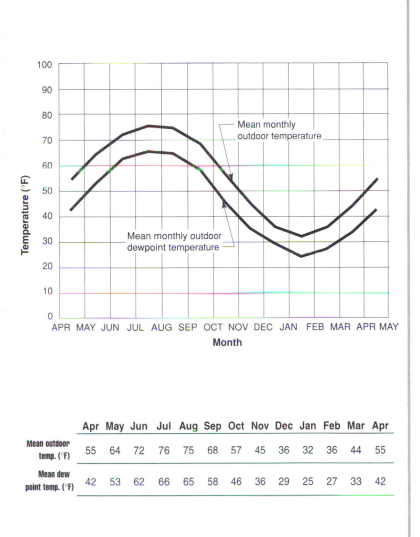

	Apr	May	Jun	Jul	Aug	Sep	Oct	Nov	Dec	Jan	Feb	Mar	Apr
Mean outdoor temp. (°F)	55	64	72	76	75	68	57	45	36	32	36	44	55
Mean dew point temp. (°F)	42	53	62	66	65	58	46	36	29	25	27	33	42

Figure I.14
Lexington, Kentucky
- Average annual precipitation 45 inches
- 4,683 heating degree days; 1,140 cooling degree days
- Winter design temperature 8°F
- Summer design temperature 91°F dry bulb, 73°F wet bulb
- Average deep ground temperature 55°F
- Latitude 38.02°

Appendices

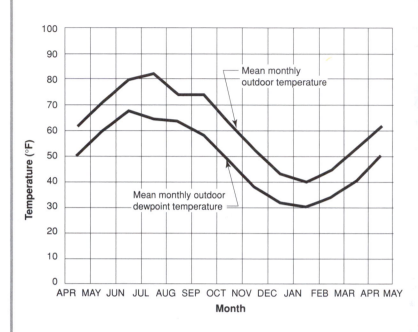

	Apr	May	Jun	Jul	Aug	Sep	Oct	Nov	Dec	Jan	Feb	Mar	Apr
Mean outdoor temp. (°F)	62	71	79	82	74	74	63	52	43	40	45	53	62
Mean dew point temp. (°F)	50	60	68	65	64	58	48	39	32	31	34	41	50

Figure I.15
Little Rock, Arkansas

- Average annual precipitation 51 inches
- 3,155 heating degree days; 2,005 cooling degree days
- Winter design temperature 20°F
- Summer design temperature 96°F dry bulb, 77°F wet bulb
- Average deep ground temperature 62°F
- Latitude 34.44°

Appendices

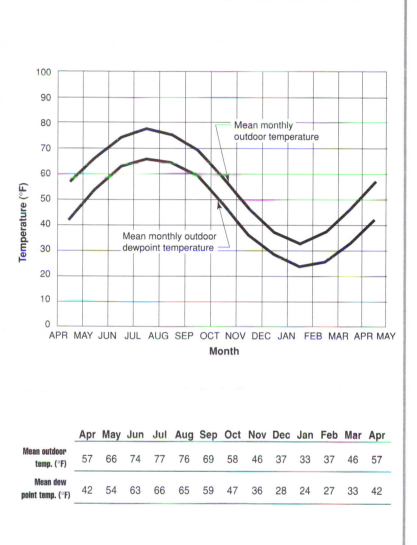

	Apr	May	Jun	Jul	Aug	Sep	Oct	Nov	Dec	Jan	Feb	Mar	Apr
Mean outdoor temp. (°F)	57	66	74	77	76	69	58	46	37	33	37	46	57
Mean dew point temp. (°F)	42	54	63	66	65	59	47	36	28	24	27	33	42

Figure I.16
Louisville, Kentucky

- Average annual precipitation 44 inches
- 4,514 heating degree days; 1,288 cooling degree days
- Winter design temperature 10°F
- Summer design temperature 93°F dry bulb, 74°F wet bulb
- Average deep ground temperature 57°F
- Latitude 38.10°

Appendices

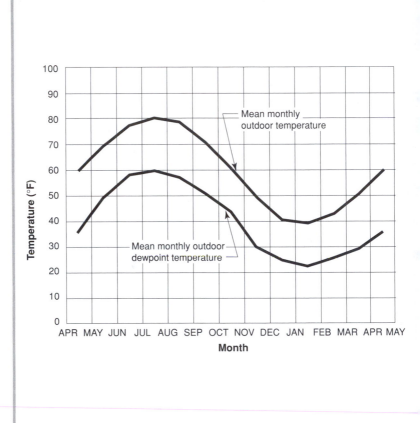

	Apr	May	Jun	Jul	Aug	Sep	Oct	Nov	Dec	Jan	Feb	Mar	Apr
Mean outdoor temp. (°F)	60	69	77	80	78	71	61	49	41	39	43	51	60
Mean dew point temp. (°F)	36	49	58	60	57	51	44	30	25	23	26	29	36

Figure I.17
Lubbock, Texas

- Average annual precipitation 19 inches
- 3,431 heating degree days; 1,689 cooling degree days
- Winter design temperature 15°F
- Summer design temperature 96°F dry bulb, 69°F wet bulb
- Average deep ground temperature 60°F
- Latitude 33.40°

12 **I**

Appendices

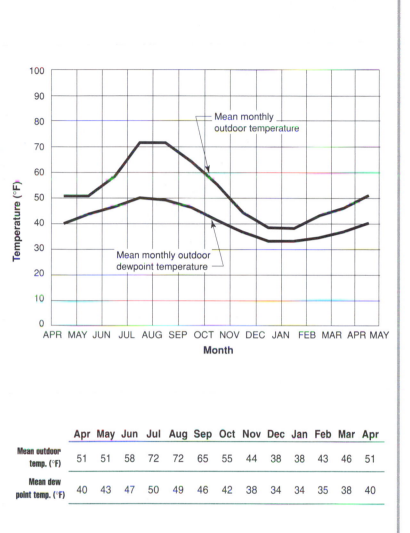

	Apr	May	Jun	Jul	Aug	Sep	Oct	Nov	Dec	Jan	Feb	Mar	Apr
Mean outdoor temp. (°F)	51	51	58	72	72	65	55	44	38	38	43	46	51
Mean dew point temp. (°F)	40	43	47	50	49	46	42	38	34	34	35	38	40

Figure I.18
Medford, Oregon
- Average annual precipitation 19 inches
- 4,611 heating degree days; 725 cooling degree days
- Winter design temperature 23°F
- Summer design temperature 94°F dry bulb, 67°F wet bulb
- Average deep ground temperature 54°F
- Latitude 42.23°

I **12**

Appendices

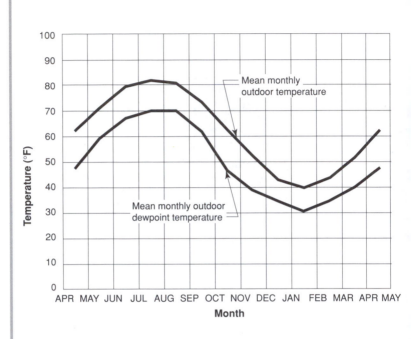

	Apr	May	Jun	Jul	Aug	Sep	Oct	Nov	Dec	Jan	Feb	Mar	Apr
Mean outdoor temp. (°F)	62	71	79	82	81	74	63	52	43	40	44	52	62
Mean dew point temp. (°F)	48	59	67	70	69	62	47	39	35	31	35	40	48

Figure I.19
Memphis, Tennessee
- Average annual precipitation 52 inches
- 3,082 heating degree days; 2,118 cooling degree days
- Winter design temperature 18°F
- Summer design temperature 95°F dry bulb, 76°F wet bulb
- Average deep ground temperature 62°F
- Latitude 35.03°

Appendices

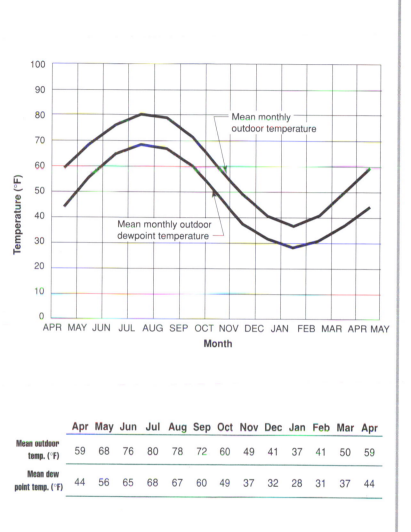

	Apr	May	Jun	Jul	Aug	Sep	Oct	Nov	Dec	Jan	Feb	Mar	Apr
Mean outdoor temp. (°F)	59	68	76	80	78	72	60	49	41	37	41	50	59
Mean dew point temp. (°F)	44	56	65	68	67	60	49	37	32	28	31	37	44

Figure I.20
Nashville, Tennessee

- Average annual precipitation 47 inches
- 3,729 heating degree days; 1,616 cooling degree days
- Winter design temperature 14°F
- Summer design temperature 94°F dry bulb, 74°F wet bulb
- Average deep ground temperature 59°F
- Latitude 36.07°

Appendices

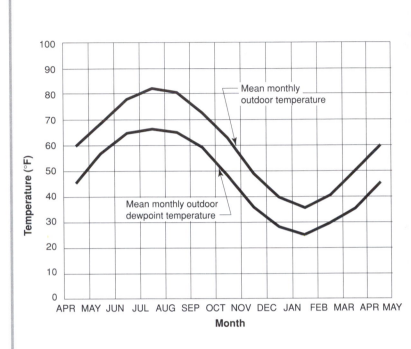

	Apr	May	Jun	Jul	Aug	Sep	Oct	Nov	Dec	Jan	Feb	Mar	Apr
Mean outdoor temp. (°F)	60	68	77	82	81	73	63	49	40	36	41	50	60
Mean dew point temp. (°F)	46	57	65	66	65	59	48	36	29	25	29	35	46

Figure I.21
Oklahoma City, Oklahoma
- Average annual precipitation 33 inches
- 3,659 heating degree days; 1,859 cooling degree days
- Winter design temperature 13°F
- Summer design temperature 97°F dry bulb, 74°F wet bulb
- Average deep ground temperature 60°F
- Latitude 35.23°

12 I

Appendices

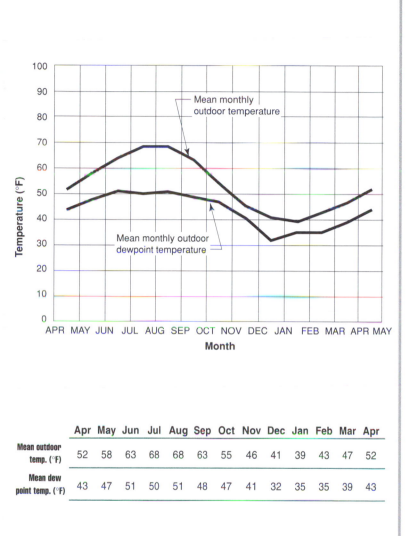

	Apr	May	Jun	Jul	Aug	Sep	Oct	Nov	Dec	Jan	Feb	Mar	Apr
Mean outdoor temp. (°F)	52	58	63	68	68	63	55	46	41	39	43	47	52
Mean dew point temp. (°F)	43	47	51	50	51	48	47	41	32	35	35	39	43

Figure I.22
Portland, Oregon

- Average annual precipitation 36 inches
- 4,522 heating degree days; 371 cooling degree days
- Winter design temperature 23°F
- Summer design temperature 85°F dry bulb, 67°F wet bulb
- Average deep ground temperature 54°F
- Latitude 45.35°

Appendices

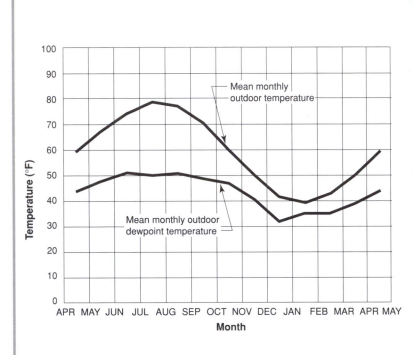

	Apr	May	Jun	Jul	Aug	Sep	Oct	Nov	Dec	Jan	Feb	Mar	Apr
Mean outdoor temp. (°F)	59	67	74	78	77	71	60	50	42	40	43	50	59
Mean dew point temp. (°F)	44	51	64	64	63	57	51	40	29	28	31	36	44

Figure I.23
Raleigh/Durham, North Carolina

- Average annual precipitation 41 inches
- 3,659 heating degree days; 1,859 cooling degree days
- Winter design temperature 20°F
- Summer design temperature 92°F dry bulb, 75°F wet bulb
- Average deep ground temperature 59°F
- Latitude 35.52°

Appendices

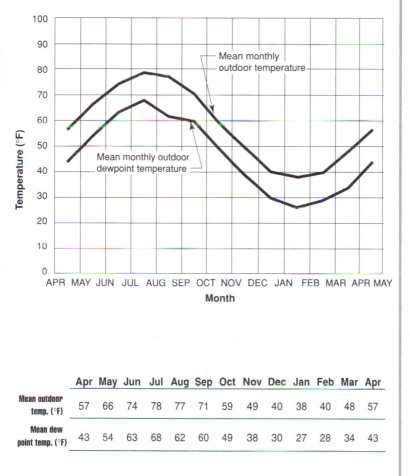

	Apr	May	Jun	Jul	Aug	Sep	Oct	Nov	Dec	Jan	Feb	Mar	Apr
Mean outdoor temp. (°F)	57	66	74	78	77	71	59	49	40	38	40	48	57
Mean dew point temp. (°F)	43	54	63	68	62	60	49	38	30	27	28	34	43

Figure I.24
Richmond, Virginia
- Average annual precipitation 43 inches
- 3,963 heating degree days; 1,348 cooling degree days
- Winter design temperature 17°F
- Summer design temperature 92°F dry bulb, 76°F wet bulb
- Average deep ground temperature 58°F
- Latitude 37.30°

Appendices

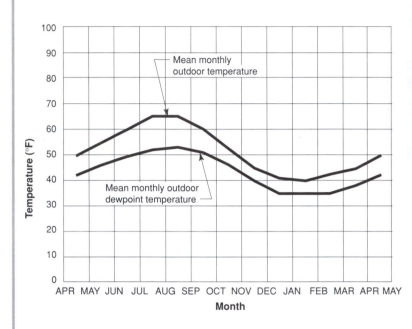

	Apr	May	Jun	Jul	Aug	Sep	Oct	Nov	Dec	Jan	Feb	Mar	Apr
Mean outdoor temp. (°F)	49	55	60	65	65	60	52	45	41	40	43	45	49
Mean dew point temp. (°F)	42	46	49	52	53	51	46	40	35	35	35	38	42

Figure I.25
Seattle/Tacoma, Washington

- Average annual precipitation 40 inches
- 4,424 heating degree days; 190 cooling degree days
- Winter design temperature 26°F
- Summer design temperature 80°F dry bulb, 64°F wet bulb
- Average deep ground temperature 52°F
- Latitude 47.27°

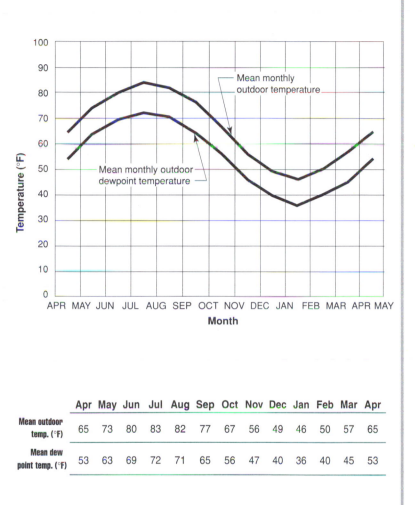

	Apr	May	Jun	Jul	Aug	Sep	Oct	Nov	Dec	Jan	Feb	Mar	Apr
Mean outdoor temp. (°F)	65	73	80	83	82	77	67	56	49	46	50	57	65
Mean dew point temp. (°F)	53	63	69	72	71	65	56	47	40	36	40	45	53

Figure I.26
Shreveport, Louisiana

- Average annual precipitation 46 inches
- 2,264 heating degree days; 2,368 cooling degree days
- Winter design temperature 25°F
- Summer design temperature 96°F dry bulb, 76°F wet bulb
- Average deep ground temperature 65°F
- Latitude 32.26°

Appendices

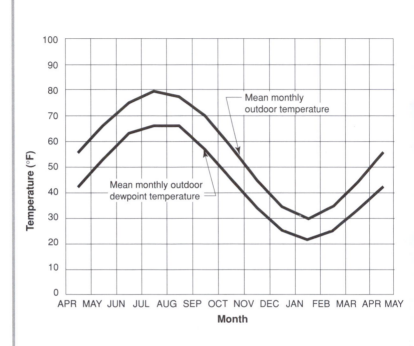

	Apr	May	Jun	Jul	Aug	Sep	Oct	Nov	Dec	Jan	Feb	Mar	Apr
Mean outdoor temp. (°F)	56	66	75	79	77	70	58	45	35	30	35	44	56
Mean dew point temp. (°F)	42	53	63	66	66	57	46	34	26	22	25	33	42

Figure I.27
St. Louis, Missouri

- Average annual precipitation 38 inches
- 4,758 heating degree days; 1,534 cooling degree days
- Winter design temperature 6°F
- Summer design temperature 94°F dry bulb, 75°F wet bulb
- Average deep ground temperature 56°F
- Latitude 38.45°

Appendices

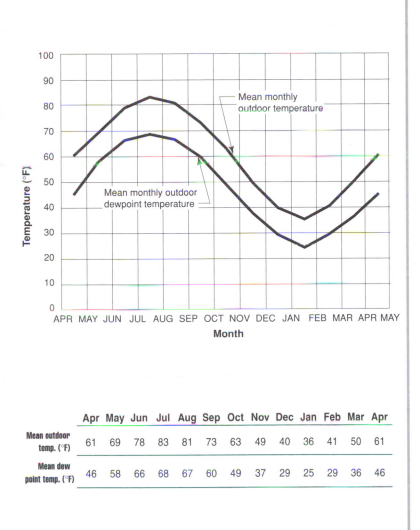

	Apr	May	Jun	Jul	Aug	Sep	Oct	Nov	Dec	Jan	Feb	Mar	Apr
Mean outdoor temp. (°F)	61	69	78	83	81	73	63	49	40	36	41	50	61
Mean dew point temp. (°F)	46	58	66	68	67	60	49	37	29	25	29	36	46

Figure I.28
Tulsa, Oklahoma
- Average annual precipitation 41 inches
- 3,691 heating degree days; 2,017 cooling degree days
- Winter design temperature 13°F
- Summer design temperature 98°F dry bulb, 75°F wet bulb
- Average deep ground temperature 60°F
- Latitude 36.11°

Appendices

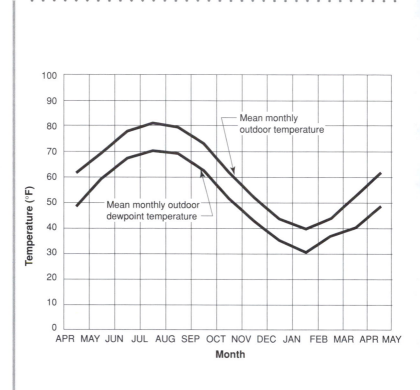

	Apr	May	Jun	Jul	Aug	Sep	Oct	Nov	Dec	Jan	Feb	Mar	Apr
Mean outdoor temp. (°F)	62	69	77	81	79	73	62	52	44	40	44	53	62
Mean dew point temp. (°F)	48	59	67	70	69	63	52	43	36	31	36	41	48

Figure I.29
Tupelo, Mississippi
- Average annual precipitation 56 inches
- 3,079 heating degree days; 1,908 cooling degree days
- Winter design temperature 19°F
- Summer design temperature 94°F dry bulb, 77°F wet bulb
- Average deep ground temperature 62°F
- Latitude 34.15°

12 **I**

Appendices

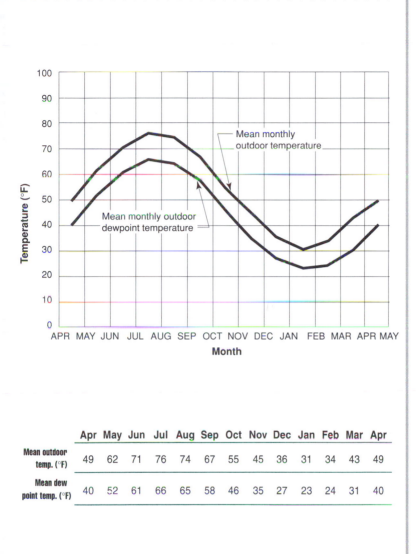

	Apr	May	Jun	Jul	Aug	Sep	Oct	Nov	Dec	Jan	Feb	Mar	Apr
Mean outdoor temp. (°F)	49	62	71	76	74	67	55	45	36	31	34	43	49
Mean dew point temp. (°F)	40	52	61	66	65	58	46	35	27	23	24	31	40

Figure I.30
Washington, DC
- Average annual precipitation 40 inches
- 4,224 heating degree days; 973 cooling degree days
- Winter design temperature 17°F
- Summer design temperature 91°F dry bulb, 74°F wet bulb
- Average deep ground temperature 54°F
- Latitude 38.56°

I 12

Appendices

Appendix II
Rain and Drainage Planes

Rain is the single most important factor to control in order to construct a durable structure. Although controlling rain has preoccupied builders for thousands of years, significant insight into the physics of rain and its control was not developed until the middle of this century by the Norwegians and the Canadians. Both peoples are blessed by countries with miserable climates which no doubt made the issue pressing.

Experience from tradition based practices combined with the physics of rain developed by the Norwegians and Canadians has provided us with effective strategies to control rain entry. The strategies are varied based on the frequency and severity of rain.

The amount of rain determines the amount of rain control needed. No rain, no rain control needed. Little rain, little rain control needed. Lots of rain, lots of rain control needed. Although this should be obvious, it is often overlooked by codes, designers, and builders. Strategies which work in Las Vegas do not necessarily work in Seattle. In simple terms, the amount of rainwater deposited on a surface determines the type of approach necessary to control rain.

Wind strength, wind direction, and rainfall intensity determine in a general way the amount of wind-driven rain deposited. These are factors governed by climate, not by design and construction. The actual distribution of rain on a building is determined by the pattern of wind flow around buildings. This, to a limited extent, can be influenced by design and construction.

Once rain is deposited on a building surface, its flow over the building surface will be determined by gravity, wind flow over the surface, and wall-surface features such as overhangs, flashings, sills, copings, and mullions. Gravity cannot be influenced by design and construction, and

II 12

Appendices

wind flow over building surfaces can only be influenced marginally. However, wall-surface features are completely within the control of the designer and builder. Tradition based practice has a legacy of developing architectural detailing features that have been used to direct water along particular paths or to cause it to drip free of the wall. Overhangs were developed for a reason. Flashings with rigid drip edges protruding from building faces were specified for a reason. Extended window sills were installed for a reason.

Rain penetration into and through building surfaces is governed by capillarity, momentum, surface tension, gravity, and wind (air pressure) forces. Capillary forces draw rainwater into pores and tiny cracks, while the remaining forces direct rainwater into larger openings.

In practice, capillarity can be controlled by capillary breaks, capillary resistant materials or by providing a receptor for capillary moisture. Momentum can be controlled by eliminating openings that go straight through the wall assembly. Rain entry by surface tension can be controlled by the use of drip edges and kerfs. Flashings and layering the wall assembly elements to drain water to the exterior (providing a "drainage plane") can be used to control rainwater from entering by gravity flow, along with simultaneously satisfying the requirements for control of momentum and surface tension forces. Sufficiently overlapping the wall assembly elements or layers comprising the drainage plane can also control entry of rainwater by air pressure differences. Finally, locating a pressure equalized air space or pressure moderated air space immediately behind the exterior cladding can be used to control entry of rainwater by air pressure differences by reducing those air pressure differences.

Coupling a pressure equalized or pressure moderated air space with a capillary resistant drainage plane represents the state-of-the-art for Norwegian and Canadian rain control practices. This approach addresses all of the driving forces responsible for rain penetration into and through building surfaces under the severest exposures.

This understanding of the physics of rain leads to the following general approach to rain control:

- reduce the amount of rainwater deposited and flowing on building surfaces

- control rainwater deposited and flowing on building surfaces

The first part of the general approach to rain control involves locating buildings so that they are sheltered from prevailing winds, providing roof overhangs and massing features to shelter exterior walls and reduce wind flow over building surfaces, and finally, providing architectural detailing to shed rainwater from building faces.

The second part of the general approach to rain control involves dealing with capillarity, momentum, surface tension, gravity and air pressure forces acting on rainwater deposited on building surfaces.

The second part of the general approach to rain control employs two general design principles:

- Face-Sealed/Barrier Approach
    Storage/Reservoir Systems
        (all rain exposures)
    Non-Storage/Non-Reservoir Systems
        (less than 20 inches average annual precipitation)

- Water Managed Approach
    Drain-Screen Systems
        (less than 40 inches average annual precipitation)
    Rain-Screen Systems
        (less than 60 inches average annual precipitation)
    Pressure Equalized Rain-Screen (PER) or Pressure
    Moderated Screen (PMS) Systems
        (all rain exposures)

Rain is permitted to enter through the cladding skin in the three water managed systems: drain-screen, rain-screen or pressure equalized rain-screen (PER)/pressure moderated screen (PMS) systems. "Drain the rain" is the cornerstone of water managed systems. In the three water managed systems, drainage of water is provided by a capillary resistant drainage plane or a capillary resistant drainage plane coupled with an air space behind the cladding. If the air space has sufficient venting to the exterior to equalize or moderate the pressure difference between the exterior and the cavity, the system is classified as a PER or PMS design.

In the face-sealed barrier approach, the exterior face is the only means to control rain entry. In storage/reservoir systems, some rain is permitted to enter and is stored in the mass of the wall assembly until drying occurs to either the exterior or interior. In non-storage/non-reservoir systems, no rain can be permitted to enter.

The performance of a specific system is determined by frequency of rain, severity of rain, system design, selection of materials, workmanship, and maintenance. In general, water managed systems out perform face-sealed/barrier systems due to their more forgiving nature. However, face-sealed/barrier systems constructed from water resistant materials that employ significant storage have a long historical track-record of exemplary performance even in the most severe rain exposures. These "massive" wall assemblies constructed out of masonry, limestone, granite and concrete, many of which are 18 inches or more thick, were typically used in public buildings such as courthouses, libraries, schools and hospitals.

The least forgiving and least water resistant assembly is a face-sealed/barrier wall constructed from water sensitive materials that does not have storage capacity. Most external insulation finish systems (EIFS) are of this type and are not generally recommended. They should be limited to climate zones which see little rain (less than 20 inches average annual precipitation).

The most forgiving and most water resistant assembly is a pressure equalized rain screen or pressure moderated screen wall constructed from water resistant materials. These types of assemblies perform well in the most severe rain exposures (more than 60 inches average annual precipitation).

Water managed strategies are recommended in all climate regions and are essential where average annual rainfall exceeds 20 inches. Drain-screen systems (drainage planes without drainage spaces) should be limited to regions where average annual rainfall is less than 40 inches and rain-screen systems (drainage planes with drainage spaces) should be limited to regions where average annual rainfall is less than 60 inches. Pressure equalized rain-screen systems (drainage planes with pressure equalized drainage spaces) or pressure moderated screen systems (drainage planes with pressure moderated drainage spaces) should be used wherever average annual rainfall is greater than 60 inches.

Face-sealed/barrier strategies should be carefully considered. Non-storage/non-reservoir systems constructed out of water sensitive materials are not generally recommended and if used should be limited to regions where average annual rainfall is less than 20 inches. Storage/reservoir systems constructed with water resistant materials can be built anywhere. However, their performance is design, workmanship, and materials dependent. In general, these systems should be limited to regions or to designs with high drying potentials to the exterior, interior or, better still, to both.

Drainage Plane Continuity

The most common residential approach to rain control is the use of a drainage plane. This drainage plane is typically a "tar paper" or building paper. More recently, the term "housewrap" has been introduced to describe building papers that are not asphalt impregnated felts (tar papers). Drainage planes can also be created by sealing or layering water resistant sheathings such as a rigid insulation or a foil covered structural sheathing.

In order to effectively "drain the rain," the drainage plane must provide drainage plane continuity especially at "punched openings" such as windows and doors. Other critical areas for drainage plane continuity are where roofs and decks intersect walls, and where piping, ducts, louvers, wiring and other services pass through the building envelope.

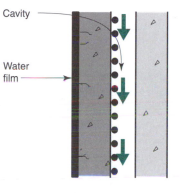

Capillary suction draws
water into porous material
and tiny cracks

Cavity acts as capillary
break and receptor for
capillary water interrupting flow

Figure II.1
Capillarity as a Driving Force for Rain Entry
- Capillary suction draws water into porous material and tiny cracks.
- Cavity acts as capillary break and receptor for capillary water interrupting flow.

II 12

Appendices

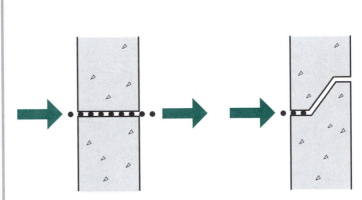

Rain droplets can be
carried through a wall
by their own momentum

Rain entry by momentum
can be prevented by designing
wall systems with no straight
through openings

Appendices

Figure II.2
Momentum as a Driving Force for Rain Entry
- Rain droplets can be carried through a wall by their own momentum.
- Rain entry by momentum can be prevented by designing wall systems with no straight through openings.

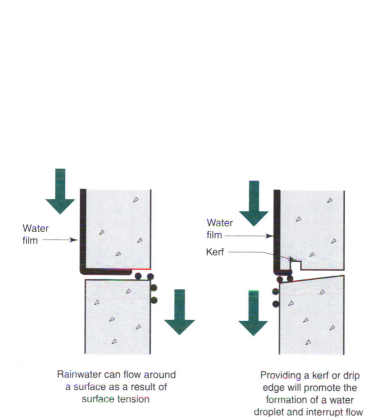

Rainwater can flow around
a surface as a result of
surface tension

Providing a kerf or drip
edge will promote the
formation of a water
droplet and interrupt flow

Figure II.3
Surface Tension as a Driving Force for Rain Entry
- Rainwater can flow around a surface as a result of surface tension.
- Providing a kerf or drip edge will promote the formation of a water droplet and interrupt flow.

II **12**

Appendices

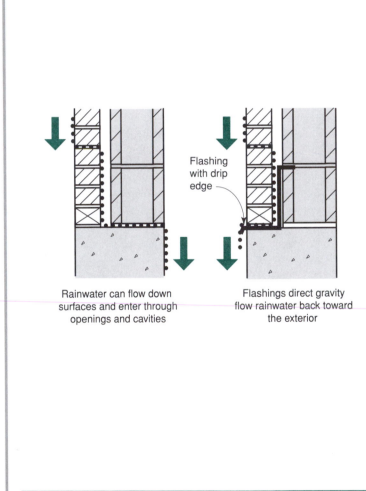

Flashing
with drip
edge

Rainwater can flow down
surfaces and enter through
openings and cavities

Flashings direct gravity
flow rainwater back toward
the exterior

12 II

Appendices

Figure II.4
Gravity as a Driving Force for Rain Entry
- Rainwater can flow down surfaces and enter through openings and cavities.
- Flashings direct gravity flow rainwater back toward the exterior.

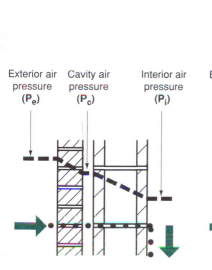

Exterior air pressure (P_e) Cavity air pressure (P_c) Interior air pressure (P_i)

$$P_e > P_c > P_i$$

Driven by air pressure differences, rain droplets are drawn through wall openings from the exterior to the interior

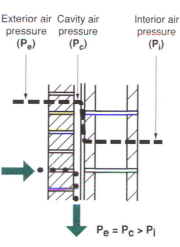

Exterior air pressure (P_e) Cavity air pressure (P_c) Interior air pressure (P_i)

$$P_e = P_c > P_i$$

By creating pressure equalization or pressure moderation between the exterior and cavity air, air pressure is diminished as a driving force for rain entry.

Figure II.5
Air Pressure Difference as a Driving Force for Rain Entry

- Driven by air pressure differences, rain droplets are drawn through wall openings from the exterior to the interior.
- By creating pressure equalization or pressure moderation between the exterior and cavity air, air pressure is diminished as a driving force for rain entry.

II **12**

Appendices

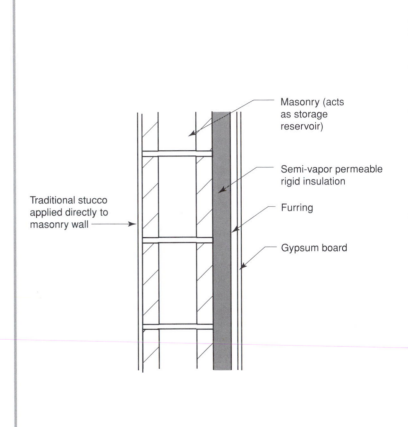

Masonry (acts as storage reservoir)

Semi-vapor permeable rigid insulation

Furring

Gypsum board

Traditional stucco applied directly to masonry wall

**Figure II.6
Face-Sealed Barrier Wall
Storage Reservoir System**

- Some rain entry past exterior face permitted.
- Penetrating rain stored in mass of wall until drying occurs to interior or exterior

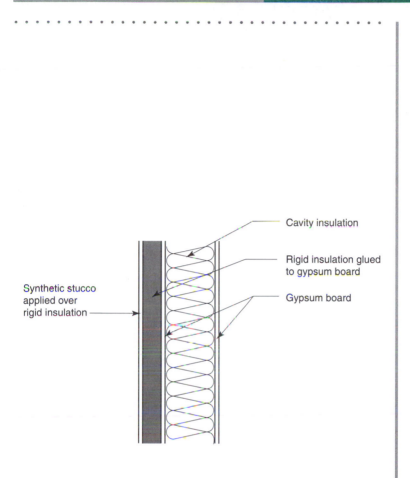

Cavity insulation

Rigid insulation glued
to gypsum board

Gypsum board

Synthetic stucco
applied over
rigid insulation

Figure II.7
Face-Sealed Barrier Wall
Non-Storage Non-Reservoir System
- No rain entry past exterior face permitted
- Should be limited to regions with a "low" rain exposure
- Should not be used in regions where the average annual precipitation
  exceeds 20 inches
- Not generally recommended

II 12

Appendices

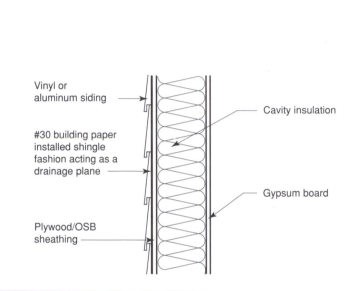

Vinyl or aluminum siding

Cavity insulation

#30 building paper installed shingle fashion acting as a drainage plane

Gypsum board

Plywood/OSB sheathing

12 II

Appendices

Figure II.8
Water Managed Wall
Drain-Screen System (Drainage Plane)
- Should be limited to regions with a "moderate" or "low" rain exposure.
- Should be used in regions where the average annual precipitation is less than 40 inches.

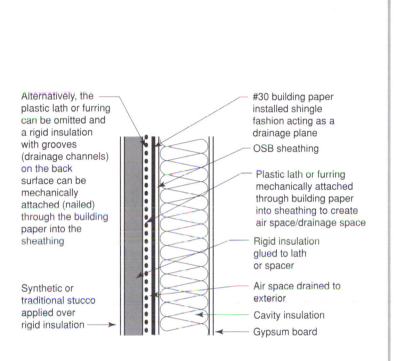

Alternatively, the plastic lath or furring can be omitted and a rigid insulation with grooves (drainage channels) on the back surface can be mechanically attached (nailed) through the building paper into the sheathing

Synthetic or traditional stucco applied over rigid insulation

#30 building paper installed shingle fashion acting as a drainage plane

OSB sheathing

Plastic lath or furring mechanically attached through building paper into sheathing to create air space/drainage space

Rigid insulation glued to lath or spacer

Air space drained to exterior

Cavity insulation

Gypsum board

Figure II.9
Water Managed Wall
Rain-Screen System (Drainage Plane with Drainage Space)
- Should be limited to regions with a "high", "moderate" or "low" rain exposure.
- Should be used in regions where the average annual precipitation exceeds 40 inches and is less than 60 inches
- To convert this wall system to a PER or PMS system, the drainage space needs to be vented to the exterior rather than simply drained

Appendices

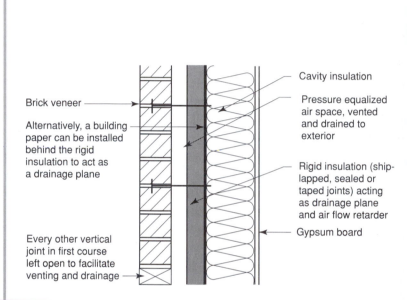

Brick veneer

Alternatively, a building paper can be installed behind the rigid insulation to act as a drainage plane

Every other vertical joint in first course left open to facilitate venting and drainage

Cavity insulation

Pressure equalized air space, vented and drained to exterior

Rigid insulation (ship-lapped, sealed or taped joints) acting as drainage plane and air flow retarder

Gypsum board

**Figure II.10
Water Managed Wall
Pressure Equalized Rain-Screen System/Pressure Moderated Screen System
(Drainage Plane with Pressure Equalized Drainage Space or Pressure Moderated Drainage Space)**

- Should be used in regions with "extreme" rain exposure
- Should be used in regions where the average annual precipitation exceeds 60 inches
- Building papers used as drainage planes are typically more reliable than taped, sealed or shiplapped rigid insulation

12 **II**

Appendices

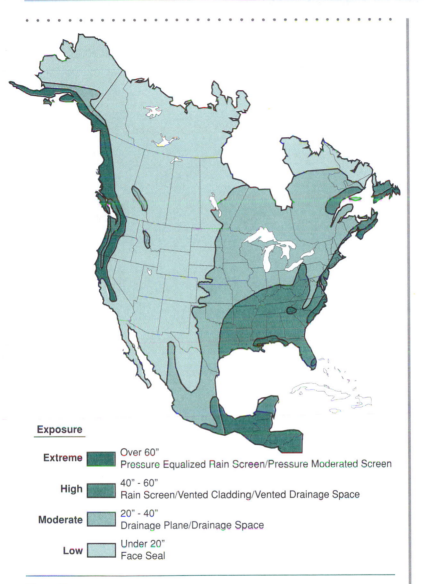

Exposure

Extreme	■	Over 60" Pressure Equalized Rain Screen/Pressure Moderated Screen
High	■	40" - 60" Rain Screen/Vented Cladding/Vented Drainage Space
Moderate	■	20" - 40" Drainage Plane/Drainage Space
Low	□	Under 20" Face Seal

Figure II.11
Annual Rainfall Map

- Based on information from the U.S. Department of Agriculture
- An example of a Drainage Plane is building paper (tar paper) installed shingle fashion
- An example of a Rain-Screen is building paper installed with a drainage space
- An example of a Pressure Equalized Rain-Screen or Pressure Moderated Screen is building paper installed with a vented drainage space
- An example of a Face-Seal is a non-drained EIFS

II **12**

Appendices

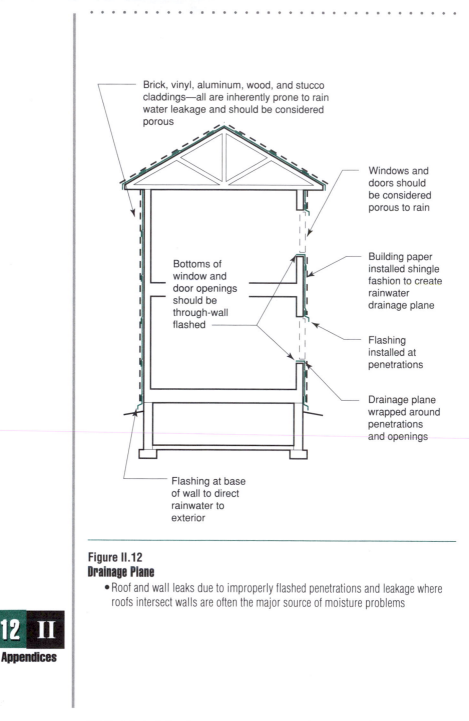

Figure II.12
Drainage Plane

- Roof and wall leaks due to improperly flashed penetrations and leakage where roofs intersect walls are often the major source of moisture problems

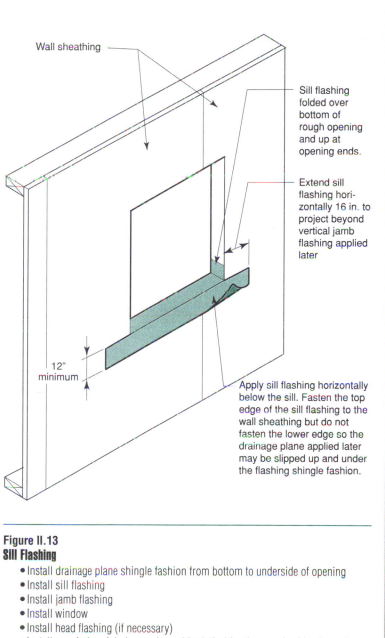

Wall sheathing

Sill flashing folded over bottom of rough opening and up at opening ends.

Extend sill flashing horizontally 16 in. to project beyond vertical jamb flashing applied later

12"
minimum

Apply sill flashing horizontally below the sill. Fasten the top edge of the sill flashing to the wall sheathing but do not fasten the lower edge so the drainage plane applied later may be slipped up and under the flashing shingle fashion.

Figure II.13
Sill Flashing

- Install drainage plane shingle fashion from bottom to underside of opening
- Install sill flashing
- Install jamb flashing
- Install window
- Install head flashing (if necessary)
- Install remainder of drainage plane shingle fashion from underside of opening up wall over side window flange
- Windows/doors with flanges typically installed in bed of sealant between flanges and jamb and sill flashing

II 12

Appendices

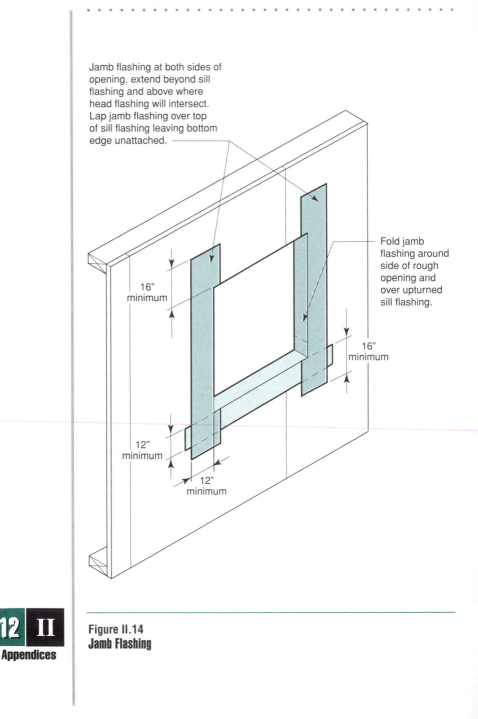

Jamb flashing at both sides of opening, extend beyond sill flashing and above where head flashing will intersect. Lap jamb flashing over top of sill flashing leaving bottom edge unattached.

Fold jamb flashing around side of rough opening and over upturned sill flashing.

16" minimum

16" minimum

12" minimum

12" minimum

**Figure II.14
Jamb Flashing**

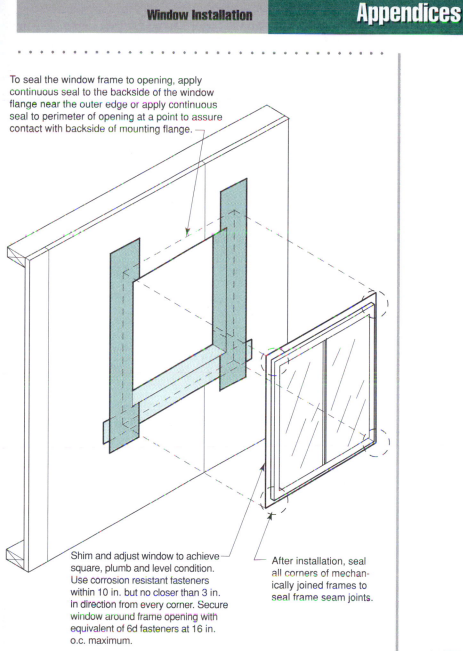

To seal the window frame to opening, apply continuous seal to the backside of the window flange near the outer edge or apply continuous seal to perimeter of opening at a point to assure contact with backside of mounting flange.

Shim and adjust window to achieve square, plumb and level condition. Use corrosion resistant fasteners within 10 in. but no closer than 3 in. in direction from every corner. Secure window around frame opening with equivalent of 6d fasteners at 16 in. o.c. maximum.

After installation, seal all corners of mechanically joined frames to seal frame seam joints.

Figure II.15
Window Installation

- Some windows are manufactured with an internal head flashing ("self-flashed"), however, these "self-flashed" windows can experience leakage at upper corners due to lack of flashing extension beyond window openings. These corners typically need additional sealing.

Appendices

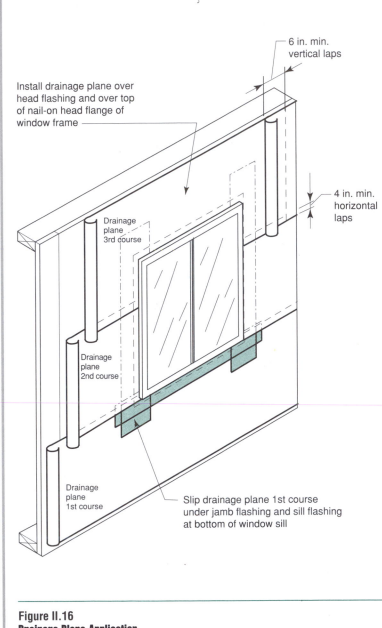

Install drainage plane over head flashing and over top of nail-on head flange of window frame

6 in. min. vertical laps

4 in. min. horizontal laps

Drainage plane 3rd course

Drainage plane 2nd course

Drainage plane 1st course

Slip drainage plane 1st course under jamb flashing and sill flashing at bottom of window sill

Figure II.16
Drainage Plane Application
- Windows/doors without flanges sealed to jamb and sill flashing round opening with sealant over backer rod

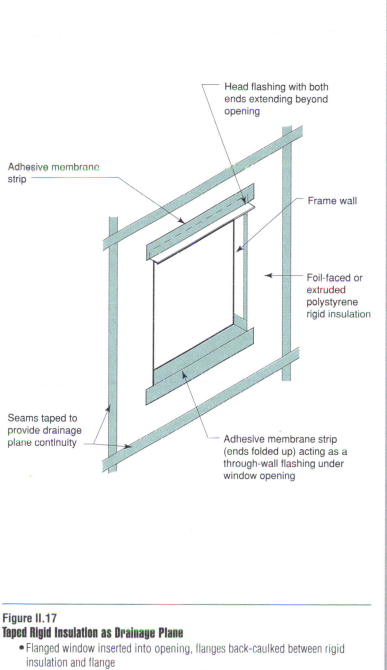

Head flashing with both ends extending beyond opening

Adhesive membrane strip

Frame wall

Foil-faced or extruded polystyrene rigid insulation

Seams taped to provide drainage plane continuity

Adhesive membrane strip (ends folded up) acting as a through-wall flashing under window opening

Figure II.17
Taped Rigid Insulation as Drainage Plane
- Flanged window inserted into opening, flanges back-caulked between rigid insulation and flange
- Tape installed over flanges sealing to rigid insulation

II **12**

Appendices

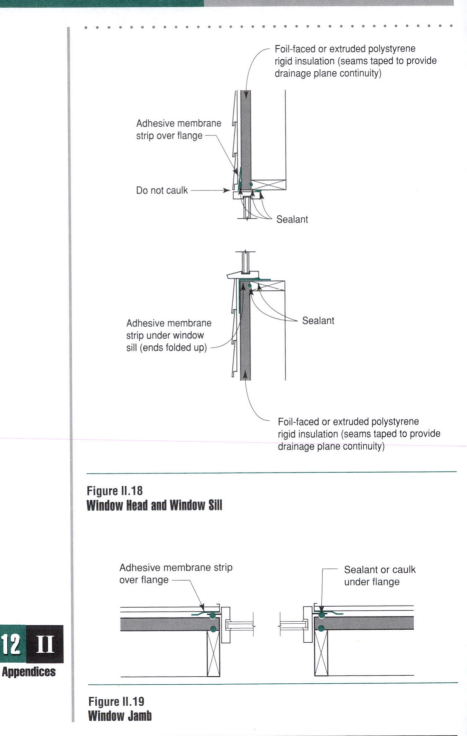

Foil-faced or extruded polystyrene
rigid insulation (seams taped to provide
drainage plane continuity)

Adhesive membrane
strip over flange

Do not caulk

Sealant

Adhesive membrane
strip under window
sill (ends folded up)

Sealant

Foil-faced or extruded polystyrene
rigid insulation (seams taped to provide
drainage plane continuity)

**Figure II.18
Window Head and Window Sill**

Adhesive membrane strip
over flange

Sealant or caulk
under flange

**Figure II.19
Window Jamb**

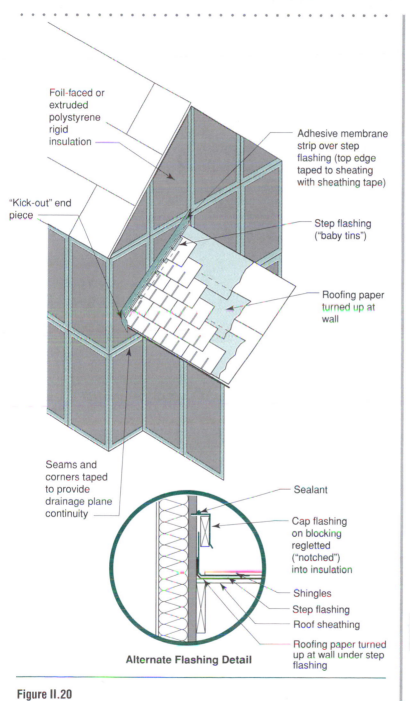

Foil-faced or extruded polystyrene rigid insulation

Adhesive membrane strip over step flashing (top edge taped to sheating with sheathing tape)

"Kick-out" end piece

Step flashing ("baby tins")

Roofing paper turned up at wall

Seams and corners taped to provide drainage plane continuity

Sealant

Cap flashing on blocking regletted ("notched") into insulation

Shingles

Step flashing

Roof sheathing

Roofing paper turned up at wall under step flashing

Alternate Flashing Detail

II 12

Figure II.20
Step Flashing

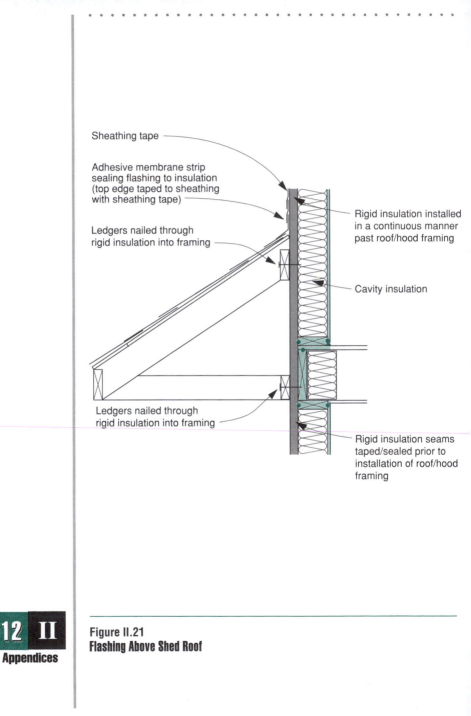

Sheathing tape

Adhesive membrane strip
sealing flashing to insulation
(top edge taped to sheathing
with sheathing tape)

Ledgers nailed through
rigid insulation into framing

Rigid insulation installed
in a continuous manner
past roof/hood framing

Cavity insulation

Ledgers nailed through
rigid insulation into framing

Rigid insulation seams
taped/sealed prior to
installation of roof/hood
framing

12 II

Appendices

Figure II.21
Flashing Above Shed Roof

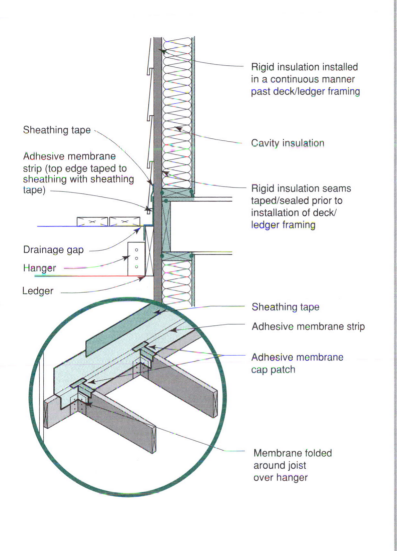

Rigid insulation installed in a continuous manner past deck/ledger framing

Sheathing tape

Adhesive membrane strip (top edge taped to sheathing with sheathing tape)

Cavity insulation

Rigid insulation seams taped/sealed prior to installation of deck/ledger framing

Drainage gap

Hanger

Ledger

Sheathing tape

Adhesive membrane strip

Adhesive membrane cap patch

Membrane folded around joist over hanger

Figure II.22
Flashing Over Deck Ledger

II **12**

Appendices

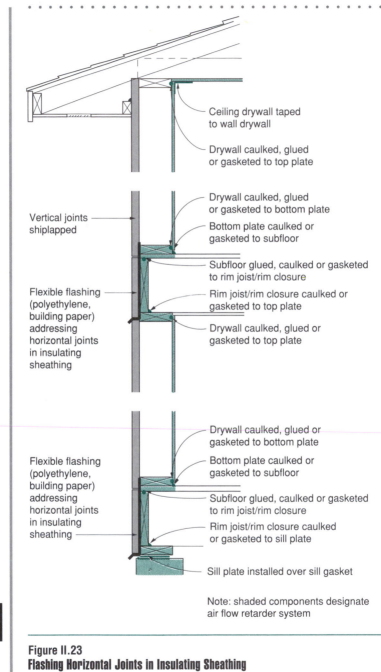

Ceiling drywall taped to wall drywall

Drywall caulked, glued or gasketed to top plate

Vertical joints shiplapped

Drywall caulked, glued or gasketed to bottom plate

Bottom plate caulked or gasketed to subfloor

Subfloor glued, caulked or gasketed to rim joist/rim closure

Flexible flashing (polyethylene, building paper) addressing horizontal joints in insulating sheathing

Rim joist/rim closure caulked or gasketed to top plate

Drywall caulked, glued or gasketed to top plate

Drywall caulked, glued or gasketed to bottom plate

Bottom plate caulked or gasketed to subfloor

Flexible flashing (polyethylene, building paper) addressing horizontal joints in insulating sheathing

Subfloor glued, caulked or gasketed to rim joist/rim closure

Rim joist/rim closure caulked or gasketed to sill plate

Sill plate installed over sill gasket

Note: shaded components designate air flow retarder system

Figure II.23
Flashing Horizontal Joints in Insulating Sheathing
- Eliminates tape at horizontal joints
- More durable than taped connection

Appendix III
Air Flow Retarders

Air flow retarders keep outside and inside air out of the building envelope. Air flow retarders can be located anywhere in the building envelope—at the exterior surface, the interior surface, or at any location in between. In practice, it is generally desirable to provide both interior and exterior air flow retarders. In heating climates, interior air flow retarders control the exfiltration of interior, often moisture-laden, air. Whereas exterior air flow retarders control the infiltration of exterior air and prevent wind-washing through insulation.

Wherever they are, air flow retarders should be:

- impermeable to air flow

- continuous over the entire building envelope

- able to withstand the forces that may act on them during and after construction

- durable over the expected lifetime of the building

Four common approaches are used to provide air flow retarders in residential buildings:

- interior air flow retarder using drywall and framing

- interior air flow retarder using polyethylene

- exterior air flow retarder using exterior sheathing

- exterior air flow retarder using building paper

Some spray applied foam insulations can be used as interstitial (cavity) air flow retarders, notably polyurethane foams. Typically applied damp spray cellulose is not an effective interstitial air flow retarder.

III 12

Appendices

An advantage of interior air flow retarders over exterior systems is that they control the entry of interior moisture-laden air into assembly cavities during heating periods. The significant disadvantage of interior air flow retarders is their inability to control wind-washing through cavity insulation.

The significant advantage of exterior air flow retarders is the ease of installation and the lack of detailing issues due to intersecting partition walls and service penetrations. However, exterior air flow retarders must deal with transitions where roof assemblies intersect exterior walls. For example, an exterior building paper ("housewrap") should be sealed to the ceiling air flow retarder system across the top of the exterior perimeter walls.

An additional advantage of exterior air flow retarder systems is the control of wind-washing that an exterior air seal provides. The significant disadvantage of exterior air flow retarders is their inability to control the entry of air-transported moisture into cavities from the interior.

Installing both interior and exterior air flow retarders addresses the weakness of each.

Air flow retarders can also be provided with properties which also class them as vapor diffusion retarders. An example of this is polyethylene film which can be used as both an air flow retarder and a vapor diffusion retarder.

Keep in mind, however, polyethylene on the inside of building assemblies in mixed-humid, mixed-dry, hot-humid and hot-dry climates is not generally a good idea (see Appendix IV); drying of building assemblies in these climates is typically to the interior (the cooler side of the assembly).

12 III

Appendices

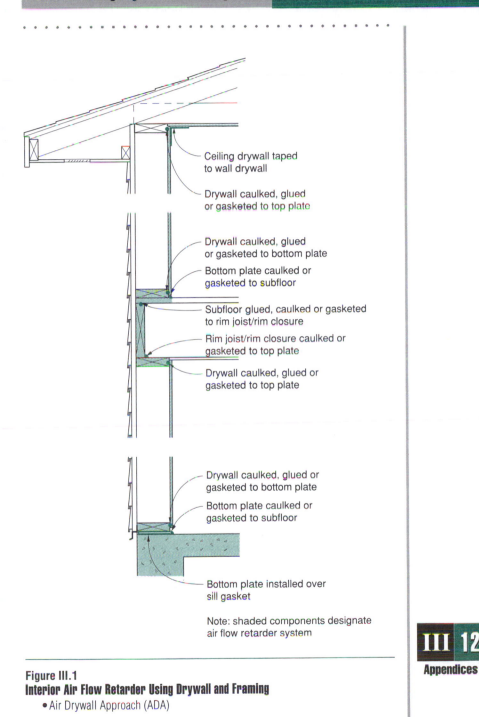

Ceiling drywall taped
to wall drywall

Drywall caulked, glued
or gasketed to top plate

Drywall caulked, glued
or gasketed to bottom plate

Bottom plate caulked or
gasketed to subfloor

Subfloor glued, caulked or gasketed
to rim joist/rim closure

Rim joist/rim closure caulked or
gasketed to top plate

Drywall caulked, glued or
gasketed to top plate

Drywall caulked, glued or
gasketed to bottom plate

Bottom plate caulked or
gasketed to subfloor

Bottom plate installed over
sill gasket

Note: shaded components designate
air flow retarder system

III 12

Appendices

Figure III.1
Interior Air Flow Retarder Using Drywall and Framing
- Air Drywall Approach (ADA)

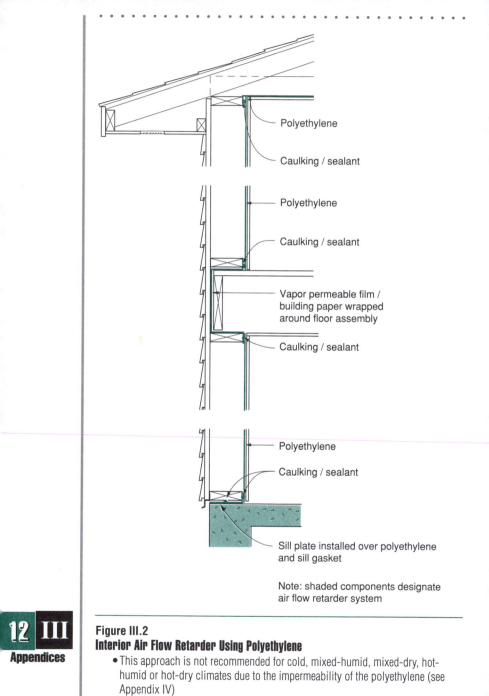

Polyethylene

Caulking / sealant

Polyethylene

Caulking / sealant

Vapor permeable film /
building paper wrapped
around floor assembly

Caulking / sealant

Polyethylene

Caulking / sealant

Sill plate installed over polyethylene
and sill gasket

Note: shaded components designate
air flow retarder system

12 III

Appendices

**Figure III.2
Interior Air Flow Retarder Using Polyethylene**
- This approach is not recommended for cold, mixed-humid, mixed-dry, hot-humid or hot-dry climates due to the impermeability of the polyethylene (see Appendix IV)
- Recommended for severe-cold climates

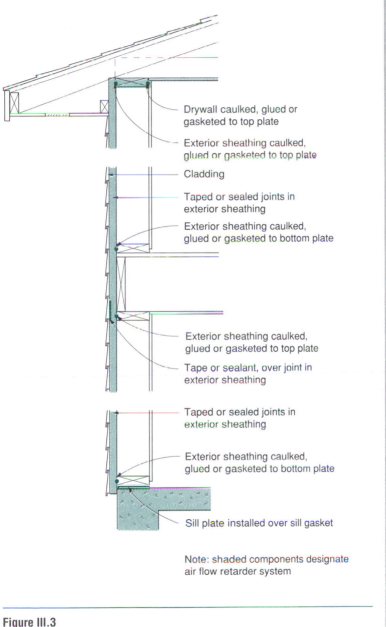

Drywall caulked, glued or gasketed to top plate

Exterior sheathing caulked, glued or gasketed to top plate

Cladding

Taped or sealed joints in exterior sheathing

Exterior sheathing caulked, glued or gasketed to bottom plate

Exterior sheathing caulked, glued or gasketed to top plate

Tape or sealant, over joint in exterior sheathing

Taped or sealed joints in exterior sheathing

Exterior sheathing caulked, glued or gasketed to bottom plate

Sill plate installed over sill gasket

Note: shaded components designate air flow retarder system

III 12

Appendices

Figure III.3
Exterior Air Flow Retarder Using Exterior Sheathing

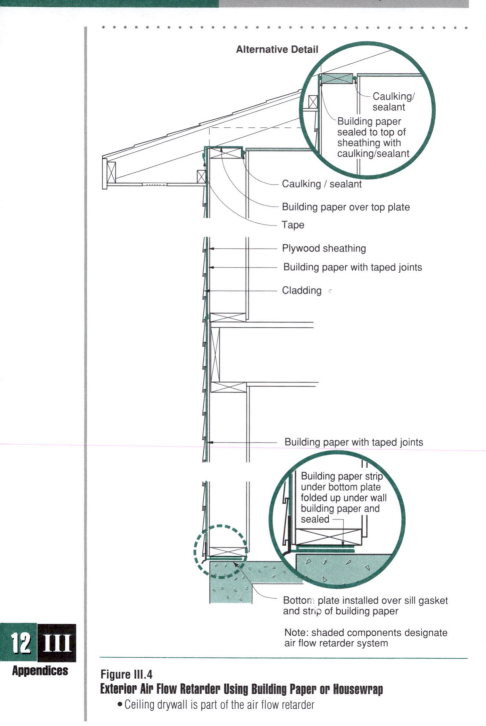

Alternative Detail

Caulking/ sealant

Building paper sealed to top of sheathing with caulking/sealant

Caulking / sealant

Building paper over top plate

Tape

Plywood sheathing

Building paper with taped joints

Cladding

Building paper with taped joints

Building paper strip under bottom plate folded up under wall building paper and sealed

Bottom plate installed over sill gasket and strip of building paper

Note: shaded components designate air flow retarder system

12 III

Appendices

Figure III.4
Exterior Air Flow Retarder Using Building Paper or Housewrap
- Ceiling drywall is part of the air flow retarder

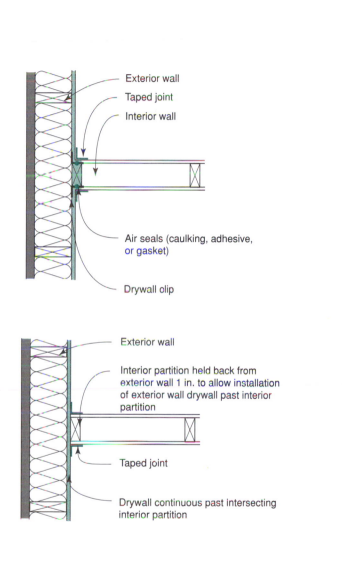

- Exterior wall
- Taped joint
- Interior wall
- Air seals (caulking, adhesive, or gasket)
- Drywall olip

- Exterior wall
- Interior partition held back from exterior wall 1 in. to allow installation of exterior wall drywall past interior partition
- Taped joint
- Drywall continuous past intersecting interior partition

Figure III.5
Intersection of Interior Partition Wall and Exterior Wall
- Air Drywall Approach (ADA)
- Seal drywall to framing at partition wall
- Alternatively, hold interior framing back and continue exterior wall drywall past intersection

III 12

Appendices

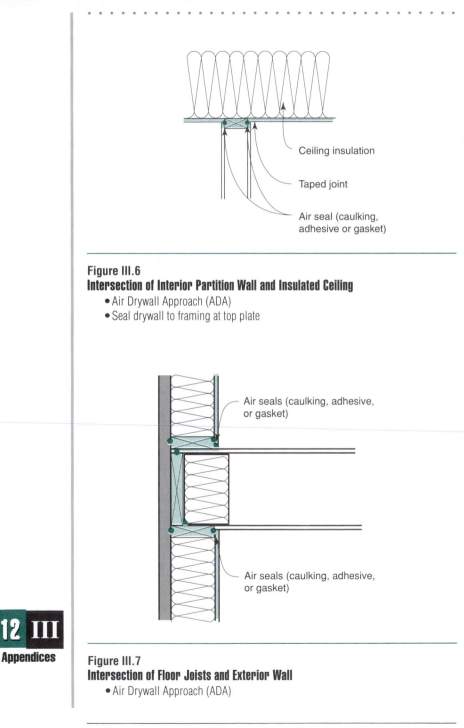

Ceiling insulation

Taped joint

Air seal (caulking, adhesive or gasket)

Figure III.6
Intersection of Interior Partition Wall and Insulated Ceiling
- Air Drywall Approach (ADA)
- Seal drywall to framing at top plate

Air seals (caulking, adhesive, or gasket)

Air seals (caulking, adhesive, or gasket)

12 III

Appendices

Figure III.7
Intersection of Floor Joists and Exterior Wall
- Air Drywall Approach (ADA)

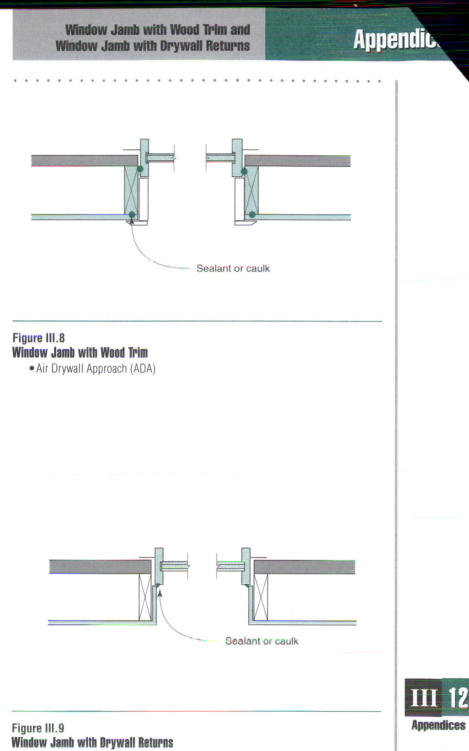

**Figure III.8
Window Jamb with Wood Trim**
- Air Drywall Approach (ADA)

**Figure III.9
Window Jamb with Drywall Returns**
- Air Drywall Approach (ADA)

Appendix IV

Insulations, Sheathings and Vapor Diffusion Retarders

Two seemingly innocuous requirements for building envelope assemblies bedevil builders and designers almost endlessly:

- keep water vapor out
- let the water vapor out if it gets in

It gets complicated because, sometimes, the best strategies to keep water vapor out also trap water vapor in. This can be a real problem if the assemblies start out wet because of rain or the use of wet materials (wet framing, concrete, masonry or damp spray cellulose, fiberglass or rock wool cavity insulation).

It gets even more complicated because of climate. In general, water vapor moves from the warm side of building assemblies to the cold side of building assemblies. This means we need different strategies for different climates. We also have to take into account differences between summer and winter.

The good news is that water vapor moves only two ways — vapor diffusion and air transport. If we understand the two ways, and know where we are (climate zone) we can solve the problem.

The bad news is that techniques that are effective at controlling vapor diffusion can be ineffective at controlling air transported moisture, and vice versa.

Building assemblies, regardless of climate zone, need to control the migration of moisture as a result of both vapor diffusion and air transport. Techniques which are effective in controlling vapor diffusion can be very different from those which control air transported moisture.

IV 12

Appendices

Vapor Diffusion and Air Transport of Vapor

Vapor diffusion is the movement of moisture in the vapor state through a material as a result of a vapor pressure difference (concentration gradient) or a temperature difference (thermal gradient). It is often confused with the movement of moisture in the vapor state into building assemblies as a result of air movement. Vapor diffusion moves moisture from an area of higher vapor pressure to an area of lower vapor pressure as well as from the warm side of an assembly to the cold side. Air transport of moisture will move moisture from an area of higher air pressure to an area of lower air pressure if moisture is contained in the moving air (Figure IV.1).

Vapor pressure is directly related to the concentration of moisture at a specific location. It also refers to the density of water molecules in air. For example, a cubic foot of air containing 2 trillion molecules of water in the vapor state has a higher vapor pressure (or higher water vapor density) than a cubic foot of air containing 1 trillion molecules of water in the vapor state. Moisture will migrate by diffusion from where there is more moisture to where there is less. Hence, moisture in the vapor state migrates by diffusion from areas of higher vapor pressure to areas of lower vapor pressure.

Moisture in the vapor state also moves from the warm side of an assembly to the cold side of an assembly. This type of moisture transport is called thermally driven diffusion. The second law of thermodynamics governs the exchange of energy and can be used to explain the concept of thermally-driven diffusion. The movement of moisture from the warm side of an assembly to the cold side of an assembly is a minimization of available "system" energy (or an increase in entropy).

When temperature differences become large, moisture vapor can condense on cold surfaces. When condensation occurs, moisture vapor is removed from the air and converted to liquid moisture on the surface resulting in a reduction in water vapor density in the air near the cold surface (i.e. a lower vapor pressure). These cold surfaces now act as "dehumidifiers" pulling more moisture towards them.

Vapor diffusion and air transport of water vapor act independently of one another. Vapor diffusion will transport moisture through materials and assemblies in the absence of an air pressure difference if a vapor pressure or temperature difference exists. Furthermore, vapor diffusion will transport moisture in the opposite direction of small air pressure differences, if an opposing vapor pressure or temperature difference exists. For example, in a hot, humid climate, the exterior is typically at a high vapor pressure and high temperature during the summer. In addition, the interior air conditioned space is maintained at a cool temperature and at a low vapor pressure through the dehumidification charac-

12 IV

Appendices

teristics of the air conditioning system, causing vapor diffusion to move water vapor from the exterior towards the interior. This will occur even if the interior conditioned space is maintained at a higher air pressure (a pressurized enclosure) relative to the exterior (Figure IV.2).

Vapor Diffusion Retarders

The function of a vapor diffusion retarder is to control the entry of water vapor into building assemblies by the mechanism of vapor diffusion. The vapor diffusion retarder may be required to control the diffusion entry of water vapor into building assemblies from the interior of a building, from the exterior of a building or from both the interior and exterior.

Vapor diffusion retarders should not be confused with air flow retarders whose function is to control the movement of air through building assemblies. In some instances, air flow retarder systems may also have specific material properties which also allow them to perform as vapor diffusion retarders. For example, a rubber membrane on the exterior of a masonry wall installed in a continuous manner is a very effective air flow retarder. The physical properties of rubber also give it the characteristics of a vapor diffusion retarder. Similarly, a continuous, sealed polyethylene ground cover installed in an unvented, conditioned crawlspace acts as both an air flow retarder and a vapor diffusion retarder. The opposite situation is also common. For example, a building paper or a housewrap installed in a continuous manner can be a very effective air flow retarder. However, the physical properties of most building papers and housewraps (they are vapor permeable - they "breathe") do not allow them to act as effective vapor diffusion retarders.

Water Vapor Permeability

The key physical property which distinguishes vapor diffusion retarders from other materials, is permeability to water vapor. Materials which retard water vapor flow are said to be impermeable. Materials which allow water vapor to pass through them are said to be permeable. However, there are degrees of impermeability and permeability and the classification of materials typically is quite arbitrary. Furthermore, under changing conditions, some materials which initially are "impermeable," can become "permeable." For example, plywood sheathing under typical conditions is relatively impermeable. However, once plywood becomes wet, it also can become relatively permeable. As a result we tend to refer to plywood as a semi-permeable material.

The unit of measurement typically used in characterizing permeability is a "perm." Many building codes define a vapor diffusion retarder as a material which has a permeability of one perm or less.

IV 12

Appendices

Materials which are generally classed as impermeable to water vapor are: rubber membranes, polyethylene film, glass, aluminum foil, sheet metal, oil-based paints, bitumen impregnated kraft paper, almost all wall coverings and their adhesives, foil-faced insulating and foil-faced non-insulating sheathings.

Materials that are generally classed as semi-permeable to water vapor are plywood, OSB, unfaced expanded polystyrene (EPS), unfaced extruded polystyrene (XPS), fiber-faced isocyanurate, heavy asphalt impregnated building papers (#30 building paper), low water-to-cement ratio concrete and most latex-based paints. Depending on the specific assembly design, construction and climate, all of these materials may or may not be considered to act as vapor diffusion retarders. Typically, these materials are considered to be more vapor permeable than vapor impermeable. Again, however, the classifications tend to be quite arbitrary.

Materials that are generally classed as permeable to water vapor are: unpainted gypsum board and plaster, unfaced fiberglass insulation, cellulose insulation, dimensional lumber and board lumber, unpainted stucco, some latex-based paints, masonry, brick, lightweight asphalt impregnated building papers (#15 building paper), asphalt impregnated fiberboard sheathings, and "housewraps."

Air Flow Retarders

The key physical properties which distinguish air flow retarders from other materials are continuity and the ability to resist air pressure differences. Continuity refers to holes, openings and penetrations. Large quantities of moisture can be transported through relatively small openings by air transport if the moving air contains moisture and if an air pressure differential also exists. For this reason, air flow retarders must be installed in such a manner that even small holes, openings and penetrations are eliminated.

Air flow retarders must also resist the air pressure differences which can act across them. These air pressure differences occur as a combination of wind, stack and mechanical system effects. Rigid materials such as interior gypsum board, exterior sheathing and rigid draftstopping materials are effective air retarders due to their ability to resist air pressure differences.

12 IV

Appendices

Magnitude of Vapor Diffusion and Air Transport of Vapor

The differences in the significance and magnitude vapor diffusion and air transported moisture are typically misunderstood. Air movement as a moisture transport mechanism is typically far more important than vapor diffusion in many (but not all) conditions. The movement of water vapor through a 1-inch-square hole as a result of a 10 Pascal air pressure differential is 100 times greater than the movement of water vapor as a result of vapor diffusion through a 32-square-foot sheet of gypsum board under normal heating or cooling conditions (See Figure IV.3).

In most climates, if the movement of moisture-laden air into a wall or building assembly is eliminated, movement of moisture by vapor diffusion is not likely to be significant. The notable exceptions are hot-humid climates or rain-wetted walls experiencing solar heating.

Furthermore, the amount of vapor which diffuses through a building component is a direct function of area. That is, if 90 percent of the building envelope area is covered with a vapor diffusion retarder, then that vapor diffusion retarder is 90 per cent effective. In other words, continuity of the vapor diffusion retarder is not as significant as the continuity of the air flow retarder. For instance, polyethylene film which may have tears and numerous punctures present will act as an effective vapor diffusion retarder, whereas at the same time it is a poor air flow retarder. Similarly, the kraft-facing on fiberglass batts installed in exterior walls acts as an effective vapor diffusion retarder, in spite of the numerous gaps and joints in the kraft-facing.

It is possible and often practical to use one material as the air flow retarder and a different material as the vapor diffusion retarder. However, the air flow retarder must be continuous and free from holes, whereas the vapor diffusion retarder need not be.

In practice, it is not possible to eliminate all holes and install a "perfect" air flow retarder. Most strategies to control air transported moisture depend on the combination of an air flow retarder, air pressure differential control and interior/exterior moisture condition control in order to be effective. Air flow retarders are often utilized to eliminate the major openings in building envelopes in order to allow the practical control of air pressure differentials. It is easier to pressurize or depressurize a building envelope made tight through the installation of an air flow retarder than a leaky building envelope. The interior moisture levels in a tight building envelope are also much easier to control by ventilation and dehumidification than those in a leaky building envelope.

Combining Approaches

In most building assemblies, various combinations of materials and approaches are often incorporated to provide for both vapor diffusion control and air transported moisture control. For example, controlling air transported moisture can be accomplished by controlling the air

IV 12

Appendices

pressure acting across a building assembly. The air pressure control is facilitated by installing an air flow retarder such as glued (or gasketed) interior gypsum board in conjunction with draftstopping. For example, in cold climates during heating periods, maintaining a slight negative air pressure within the conditioned space will control the exfiltration of interior moisture-laden air. However, this control of air transported moisture will not control the migration of water vapor as a result of vapor diffusion. Accordingly, installing a vapor diffusion retarder towards the interior of the building assembly, such as the kraft paper backing on fiberglass batts is also typically necessary. Alternatives to the kraft paper backing are low permeability paint on the interior gypsum board surfaces, the foil backing on foil-backed gypsum board, sheet polyethylene installed between the interior gypsum board and the wall framing, or almost any interior wall covering.

In the above example, control of both vapor diffusion and air transported moisture in cold climates during heating periods can be enhanced by maintaining the interior conditioned space at relatively low moisture levels through the use of controlled ventilation and source control. Also, in the above example, control of air transported moisture during cooling periods (when moisture flow is typically from the exterior towards the interior) can be facilitated by maintaining a slight positive air pressure across the building envelope thereby preventing the infiltration of exterior, hot, humid air.

Overall Strategy

Building assemblies need to be protected from wetting by air transport and vapor diffusion. The typical strategies used involve vapor diffusion retarders, air flow retarders, air pressure control, and control of interior moisture levels through ventilation and dehumidification via air conditioning. The location of air flow retarders and vapor diffusion retarders, pressurization versus depressurization, and ventilation versus dehumidification depend on climate location and season.

The overall strategy is to keep building assemblies from getting wet from the interior, from getting wet from the exterior, and allowing them to dry to either the interior or exterior should they get wet or start out wet as a result of the construction process or through the use of wet materials.

In general moisture moves from warm to cold. In cold climates, moisture from the interior conditioned spaces attempts to get to the exterior by passing through the building envelope. In hot climates, moisture from the exterior attempts to get to the cooled interior by passing through the building envelope.

Cold Climates

In cold climates and during heating periods, building assemblies need to be protected from getting wet from the interior. As such, vapor diffusion retarders and air flow retarders are installed towards the interior warm surfaces. Furthermore, conditioned spaces should be maintained at relatively low moisture levels through the use of controlled ventilation (dilution) and source control.

In cold climates the goal is to make it as difficult as possible for the building assemblies to get wet from the interior. The first line of defense is the control of moisture entry from the interior by installing interior vapor diffusion retarders, interior air flow retarders along with ventilation (dilution with exterior air) and source control to limit interior moisture levels. Since it is likely that building assemblies will get wet, a degree of forgiveness should also be designed into building assemblies allowing them to dry should they get wet. In cold climates and during heating periods, building assemblies dry towards the exterior. Therefore, permeable ("breathable") materials are often specified as exterior sheathings.

In general, in cold climates, air flow retarders and vapor diffusion retarders are installed on the interior of building assemblies, and building assemblies are allowed to dry to the exterior by installing permeable sheathings towards the exterior. A "classic" cold climate wall assembly is presented in Figure IV.4.

Hot Climates

In hot climates and during cooling periods the opposite is true. Building assemblies need to be protected from getting wet from the exterior, and allowed to dry towards the interior. Accordingly, air flow retarders and vapor diffusion retarders are installed on the exterior of building assemblies, and building assemblies are allowed to dry towards the interior by using permeable interior wall finishes, installing cavity insulations without vapor diffusion retarders (unfaced fiberglass batts) and avoiding interior "non-breathable" wall coverings such as vinyl wallpaper. Furthermore, conditioned spaces are maintained at a slight positive air pressure with conditioned (dehumidified) air in order to limit the infiltration of exterior, warm, potentially humid air (in hot-humid climates rather than hot-dry climates). A "classic" hot climate wall assembly is presented in Figure IV.5.

Mixed Climates

In mixed climates, the situation becomes more complicated. Building assemblies need to be protected from getting wet from both the interior and exterior, and be allowed to dry to either the exterior or interior.

IV 12

Appendices

Three general strategies are typically employed:

- Selecting either a classic cold climate assembly or classic hot climate assembly, using air pressure control (typically only pressurization during cooling), using interior moisture control (ventilation/air change during heating, dehumidification/air conditioning during cooling) and relying on the forgiveness of the classic approaches to dry the accumulated moisture (from opposite season exposure) to either the interior or exterior. In other words, the moisture accumulated in a cold climate wall assembly exposed to hot climate conditions is anticipated to dry towards the exterior when the cold climate assembly finally sees heating conditions, and vice versa for hot climate building assemblies;

- Adopting a "flow-through" approach by using permeable or semi-permeable building materials on both the interior and exterior surfaces of building assemblies to allow water vapor by diffusion to "flow-through" the building assembly without accumulating. Flow would be from the interior to exterior during heating periods, and from the exterior towards the interior during cooling periods. In this approach air pressure control and using interior moisture control would also occur. The location of the air flow retarder can be towards the interior (sealed interior gypsum board), or towards the exterior (sealed exterior sheathing). A "classic" flow-through wall assembly is presented in Figure IV.6; or

- Installing the vapor diffusion retarder roughly in the middle of the assembly from a thermal perspective. This is typically accomplished by installing impermeable or semi-permeable insulating sheathing on the exterior of a frame cavity wall (see Figure IV.7). For example, installing 1.5 inches of foil-faced insulating sheathing (approximately R-10) on the exterior of a 2x6 frame cavity wall insulated with unfaced fiberglass batt insulation (approximately R-19). The vapor diffusion retarder is the interior face of the exterior impermeable insulating sheathing (Figure IV.7). If the wall assembly's total thermal resistance is R-29 (R-19 plus R-10), the location of the vapor diffusion retarder is 66 percent of the way (thermally) towards the exterior (19/29 = .66). In this approach air pressure control and using interior moisture control would also occur. The location of the air flow retarder can be towards the interior or exterior.

The advantage of the wall assembly described in Figure IV.7 is that an interior vapor diffusion retarder is not necessary. In fact, locating an interior vapor diffusion retarder at this location would be detrimental, as it would not allow the wall assembly to dry towards the interior during

12 IV

Appendices

cooling periods. The wall assembly is more forgiving without the interior vapor diffusion retarder than if one were installed. If an interior vapor diffusion retarder were installed, this would result in a vapor diffusion retarder on both sides of the assembly significantly impairing durability.

Note that this discussion relates to a wall located in a mixed climate with an exterior impermeable or semi-permeable insulating sheathing. Could a similar argument be made for a cold climate wall assembly? Could we construct a wall in a cold climate without an interior vapor diffusion retarder? How about a wall in a cold climate with an exterior vapor diffusion retarder and no interior vapor diffusion retarder? The answer is yes to both questions, but with caveats.

Control of Condensing Surface Temperatures

The performance of a wall assembly in a cold climate without an interior vapor diffusion retarder (such as the wall described in Figure IV.7) can be more easily understood in terms of condensation potentials and the control of condensing surface temperatures.

Figure IV.8 illustrates the performance of a 2x6 wall with semi-permeable plywood sheathing (perm rating of about 0.5 perms, dry cup; 3.0 perms wet cup) covered with building paper and painted wood siding located in Chicago, IL. The wood siding is installed directly over the building paper, without an airspace or provision for drainage. The interior conditioned space is maintained at a relative humidity of 35 percent at 70 degrees Fahrenheit. For the purposes of this example, it is assumed that no interior vapor diffusion retarder is installed (unpainted drywall as an interior finish over unfaced fiberglass, yech!). This illustrates a case we would never want to construct in a cold climate, a wall with a vapor diffusion retarder on the exterior (semi-permeable plywood sheathing and painted wood siding without an airspace) and no vapor diffusion retarder on the interior.

The mean daily ambient temperature over a one-year period is plotted (see Figure IV.8). The temperature of the insulation/plywood sheathing interface (back side of the plywood sheathing) is approximately equivalent to the mean daily ambient temperature, since the thermal resistance values of the siding, building paper and the plywood sheathing are small compared to the thermal resistance of the insulation in the wall cavity. The dew point temperature of the interior air/water vapor mix is approximately 40 degrees Fahrenheit (this can be found from examining a psychrometric chart). In other words, whenever the back side of the plywood sheathing drops below 40 degrees Fahrenheit, the potential for condensation exists at that interface should moisture mi-

IV 12

Appendices

grate from the interior conditioned space via vapor diffusion or air movement.

From the plot it is clear that the mean daily temperature of the back side of the plywood sheathing drops below the dew point temperature of the interior air at the beginning of November and does not go above the dew point temperature until early March. The shaded area under the dew point line is the potential for condensation, or wetting potential for this assembly should moisture from the interior reach the back side of the plywood sheathing. With no interior vapor diffusion retarder, moisture from the interior will reach the back side of the plywood sheathing.

Figure IV.9 illustrates the performance of the wall assembly described in Figure IV.7, a 2x6 wall insulated on the exterior with 1.5 inches of rigid, foil-faced, impermeable, insulating sheathing (approximately R-10, perm rating of about 0.5 perms, wet cup and dry cup), located in Chicago, IL. The wall cavity is insulated with unfaced fiberglass batt insulation (approximately R-19). Unpainted drywall is again the interior finish (no interior vapor diffusion retarder). Now this wall assembly also has a vapor diffusion retarder on the exterior, but with a huge difference. This exterior vapor diffusion retarder has a significant insulating value since it is a rigid insulation. The temperature of the first condensing surface within the wall assembly, namely the cavity insulation/rigid insulation interface (the back side of the rigid insulation), is raised above the interior dew point temperature because of the insulating value of the rigid insulation. This illustrates a case we could construct in a cold climate, a wall with a "warm" vapor diffusion retarder on the exterior and no vapor diffusion retarder on the interior.

The temperature of the condensing surface (back side of the rigid insulation) is calculated in the following manner. Divide the thermal resistance to the exterior of the condensing surface by the total thermal resistance of the wall. Then multiply this ratio by the temperature difference between the interior and exterior. Finally, add this to the outside base temperature.

$$T_{interface} = R_{exterior} / R_{total} \times (T_{in} - T_{out}) + T_{out}$$

where:

$T_{interface}$ = the temperature at the sheathing/insulation interface or the temperature of the first condensing surface

$R_{exterior}$ = the R-value of the exterior sheathing

R_{total} = the total R-value of the entire wall assembly

T_{in} = the interior temperature

T_{out} = the exterior temperature

12 IV

Appendices

The R-10 insulating sheathing raises the dew point temperature at the first condensing surface so that no condensation will occur when interior moisture levels are less than 35 percent relative humidity at 70 degrees Fahrenheit. In other words, no interior vapor diffusion retarder of any kind is necessary with this wall assembly if the interior relative humidity is kept below 35 percent. This is a "caveat" for this wall assembly. Now remember, this wall is located in Chicago, IL. This is another "caveat" for this wall assembly.

What happens if we move this wall to Minneapolis? Big change. Minneapolis is a miserable place in the winter. The interior relative humidity would have to be kept below 25 percent to prevent condensation at the first condensing surface. What happens if we move the wall back to Chicago, IL, and install a modest interior vapor diffusion retarder, such as one coat of a standard interior latex paint (perm rating of about 3 perms) over the previously unpainted drywall (perm rating of 20)? If we control air leakage, interior relative humidities can be raised above 50 percent before condensation occurs.

What happens if we move this wall to Raleigh, NC, and reduce the thickness of the rigid insulation? Another big change. Raleigh has a moderate winter. Figure IV.10 illustrates the performance of a 2x6 wall insulated on the exterior with 1" of rigid, foil-faced, impermeable, insulating sheathing (approximately R-7.5, perm rating of about 0.5 perms, wet cup and dry cup), located in Raleigh, NC. In Raleigh, NC, with no interior vapor diffusion retarder of any kind, condensation will not occur until interior moisture level are raised above 45 percent, 70 degrees Fahrenheit during the coldest part of the heating season. Since these interior conditions are not likely (or desirable), the potential for condensation in this wall assembly is small.

Sheathings and Cavity Insulations

Exterior sheathings can be permeable, semi-permeable, impermeable, insulating and non-insulating. Mixing and matching sheathings, building papers and cavity insulations should be based on climate location and therefore can be challenging. The following guidelines are offered:

- Impermeable non-insulating sheathings are not recommended in cold climates (drying not possible to interior due to requirement for interior vapor diffusion retarder, condensing surface temperature not controlled due to non-insulating sheathing).

- Impermeable and semi-permeable sheathings (except plywood or OSB due to their higher permeability) are not recommended for use with damp spray cellulose cavity insulations in cold climates (drying not possible to interior due to interior vapor diffusion retarder).

- Impermeable insulating sheathings should be of sufficient thermal resistance to control condensation at cavity insulation/sheathing interfaces.

- Permeable sheathings are not recommended for use with brick veneers and stuccos due to moisture flow reversal from solar radiation (sun heats wet brick driving moisture into wall assembly through permeable sheathing).

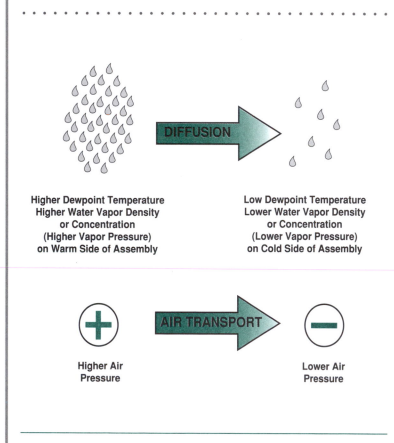

Higher Dewpoint Temperature
Higher Water Vapor Density
or Concentration
(Higher Vapor Pressure)
on Warm Side of Assembly

Low Dewpoint Temperature
Lower Water Vapor Density
or Concentration
(Lower Vapor Pressure)
on Cold Side of Assembly

Higher Air
Pressure

Lower Air
Pressure

Figure IV.1
Water Vapor Movement

- Vapor diffusion is the movement of moisture in the vapor state as a result of a vapor pressure difference (concentration gradient) or a temperature difference (thermal gradient)
- Air transport is the movement of moisture in the vapor state as a result of an air pressure difference

12 IV

Appendices

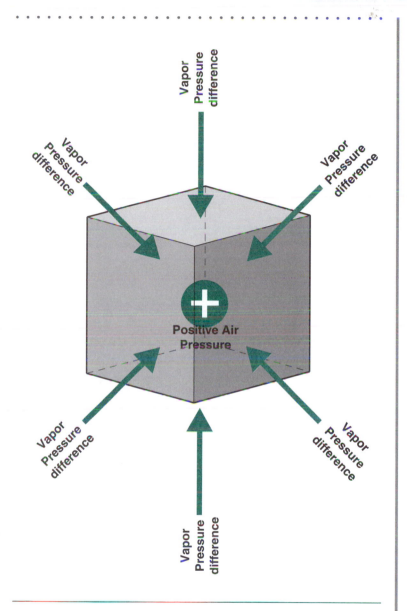

Figure IV.2
Opposing Air and Vapor Pressure Differences
- The atmosphere within the cube is under higher air pressure but lower vapor pressure relative to surroundings.
- Vapor pressure acts inward in this example.
- Air pressure acts outward in this example.

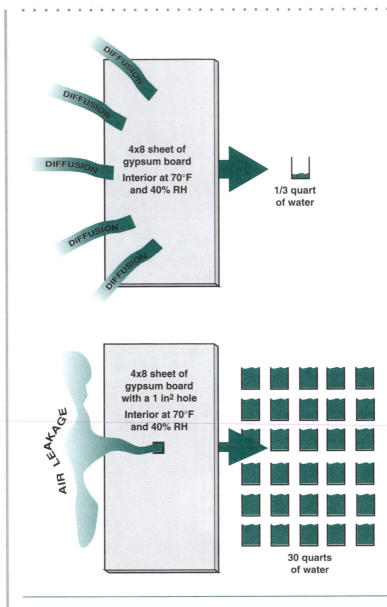

Figure IV.3
Diffusion vs. Air Leakage

- In most cold climates over an entire heating season, 1/3 of a quart of water can be collected by diffusion through gypsum board without a vapor diffusion retarder; 30 quarts of water can be collected through air leakage.

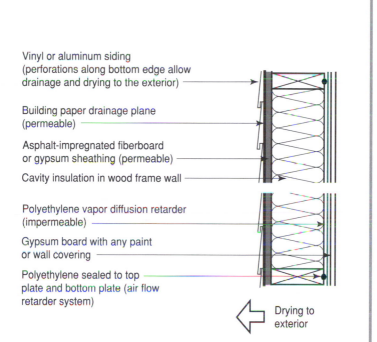

Vinyl or aluminum siding
(perforations along bottom edge allow
drainage and drying to the exterior)

Building paper drainage plane
(permeable)

Asphalt-impregnated fiberboard
or gypsum sheathing (permeable)

Cavity insulation in wood frame wall

Polyethylene vapor diffusion retarder
(impermeable)

Gypsum board with any paint
or wall covering

Polyethylene sealed to top
plate and bottom plate (air flow
retarder system)

Drying to
exterior

Figure IV.4
Classic Severe-Cold Climate Wall Assembly
- Vapor diffusion retarder to the interior
- Air flow retarder to the interior
- Permeable exterior sheathing and permeable building paper drainage plane
- Ventilation provides air change (dilution) and also limits the interior moisture
  levels.

IV 12

Appendices

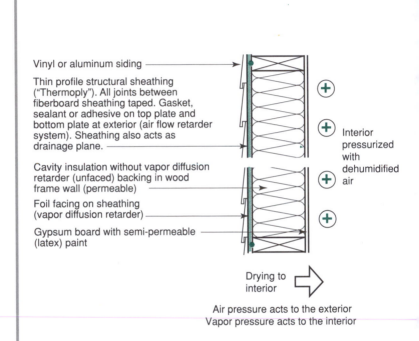

Vinyl or aluminum siding

Thin profile structural sheathing ("Thermoply"). All joints between fiberboard sheathing taped. Gasket, sealant or adhesive on top plate and bottom plate at exterior (air flow retarder system). Sheathing also acts as drainage plane.

Cavity insulation without vapor diffusion retarder (unfaced) backing in wood frame wall (permeable)

Foil facing on sheathing (vapor diffusion retarder)

Gypsum board with semi-permeable (latex) paint

Interior pressurized with dehumidified air

Drying to interior ⇨

Air pressure acts to the exterior
Vapor pressure acts to the interior

Figure IV.5
Classic Hot-Humid Climate Wall Assembly

- Vapor diffusion retarder to the exterior
- Air flow retarder to the exterior
- Pressurization of conditioned space
- Impermeable exterior sheathing also acts as drainage plane
- Permeable interior wall finish
- Interior conditioned space is maintained at a slight positive air pressure with respect to the exterior to limit the infiltration of exterior, hot, humid air
- Air conditioning also provides dehumidification (moisture removal) from interior

12 IV

Appendices

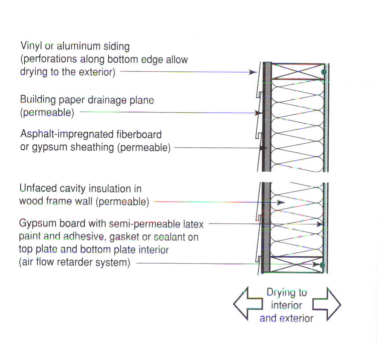

Vinyl or aluminum siding
(perforations along bottom edge allow
drying to the exterior)

Building paper drainage plane
(permeable)

Asphalt-impregnated fiberboard
or gypsum sheathing (permeable)

Unfaced cavity insulation in
wood frame wall (permeable)

Gypsum board with semi-permeable latex
paint and adhesive, gasket or sealant on
top plate and bottom plate interior
(air flow retarder system)

Drying to
interior
and exterior

Figure IV.6
Classic Flow-Through Wall Assembly

- Permeable Interior surface and finish and permeable exterior sheathing and permeable building paper drainage plane
- Interior conditioned space is maintained at a slight positive air pressure with respect to the exterior to limit the infiltration of exterior moisture-laden air during cooling
- Ventilation provides air change (dilution) and also limits the interior moisture levels during heating
- Air conditioning/dehumidification limits the interior moisture levels during cooling

IV 12

Appendices

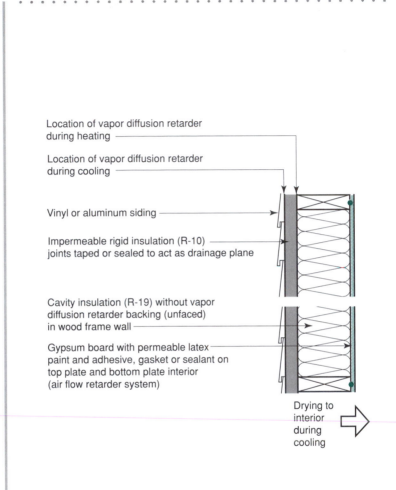

Location of vapor diffusion retarder during heating

Location of vapor diffusion retarder during cooling

Vinyl or aluminum siding

Impermeable rigid insulation (R-10) joints taped or sealed to act as drainage plane

Cavity insulation (R-19) without vapor diffusion retarder backing (unfaced) in wood frame wall

Gypsum board with permeable latex paint and adhesive, gasket or sealant on top plate and bottom plate interior (air flow retarder system)

Drying to interior during cooling

Figure IV.7
Vapor Diffusion Retarder in the Middle of the Wall

- Air flow retarder to the interior
- Permeable interior wall finish
- Interior conditioned space is maintained at a slight positive air pressure with respect to the exterior to limit the infiltration of exterior moisture-laden air during cooling.
- Ventilation provides air change (dilution) and also limits the interior moisture levels during heating.
- Air conditioning/dehumidification limits the interior moisture levels during cooling.
- Impermeable exterior sheathing also acts as drainage plane

12 IV

Appendices

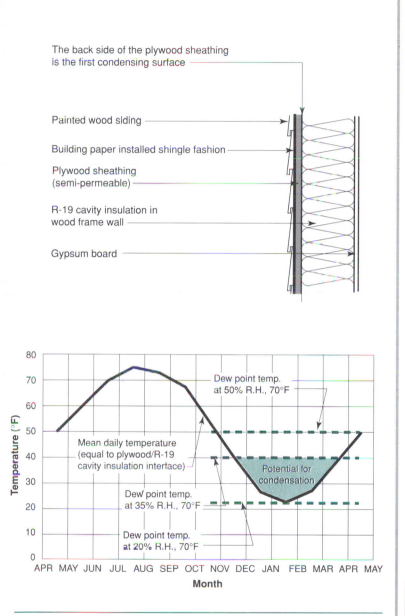

The back side of the plywood sheathing is the first condensing surface

Painted wood siding

Building paper installed shingle fashion

Plywood sheathing (semi-permeable)

R-19 cavity insulation in wood frame wall

Gypsum board

Dew point temp. at 50% R.H., 70°F

Mean daily temperature (equal to plywood/R-19 cavity insulation interface)

Potential for condensation

Dew point temp. at 35% R.H., 70°F

Dew point temp. at 20% R.H., 70°F

Month

APR MAY JUN JUL AUG SEP OCT NOV DEC JAN FEB MAR APR MAY

Figure IV.8
Potential for Condensation in a Wood Frame Wall Cavity in Chicago, IL
- By reducing interior moisture levels, the potential condensation is reduced or eliminated.

Appendices

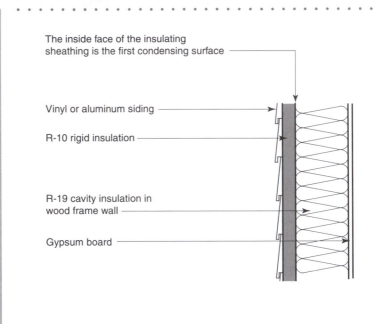

The inside face of the insulating sheathing is the first condensing surface

Vinyl or aluminum siding

R-10 rigid insulation

R-19 cavity insulation in wood frame wall

Gypsum board

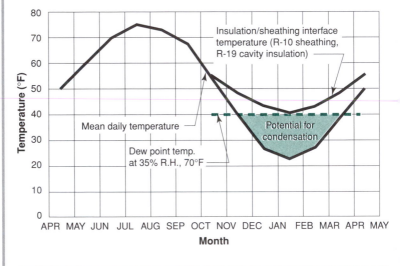

Figure IV.9

Potential for Condensation in a Wood Frame Wall Cavity Without an Interior Vapor Diffusion Retarder in Chicago, IL

- The R-10 insulating sheathing raises the dew point temperature at the first condensing surface (cavity side of the foam sheathing) so that no condensation will occur when interior moisture levels are less than 35 percent relative humidity at 70 degrees Fahrenheit.

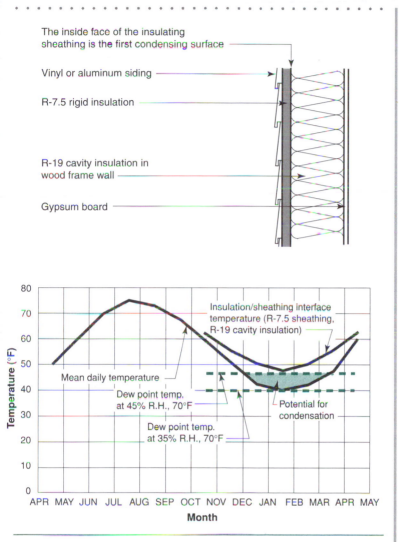

The inside face of the insulating sheathing is the first condensing surface

Vinyl or aluminum siding

R-7.5 rigid insulation

R-19 cavity insulation in wood frame wall

Gypsum board

Insulation/sheathing interface temperature (R-7.5 sheathing, R-19 cavity insulation)

Mean daily temperature

Dew point temp. at 45% R.H., 70°F

Dew point temp. at 35% R.H., 70°F

Potential for condensation

Temperature (°F)

Month

Figure IV.10
Potential for Condensation in a Wood Frame Wall Cavity Without an Interior Vapor Diffusion Retarder in Raleigh, NC

- The R-7.5 insulating sheathing raises the dew point temperature at the first condensing surface (cavity side of the foam sheathing) so that no condensation will occur when interior moisture levels are less than 45% relative humidity at 70 degrees Fahrenheit.
- Condensation will not occur in this wall assembly until interior moisture levels are raised above 45% relative humidity, 70 degrees Fahrenheit during the coldest part of the heating season. Since these interior conditions are not likely (or desirable), the potential for condensation in this wall assembly is small.

IV 12

Appendices

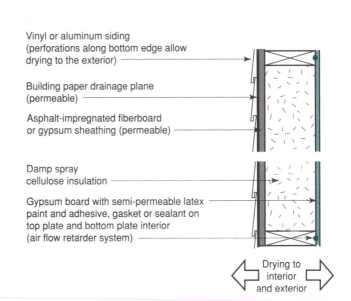

Vinyl or aluminum siding
(perforations along bottom edge allow
drying to the exterior)

Building paper drainage plane
(permeable)

Asphalt-impregnated fiberboard
or gypsum sheathing (permeable)

Damp spray
cellulose insulation

Gypsum board with semi-permeable latex
paint and adhesive, gasket or sealant on
top plate and bottom plate interior
(air flow retarder system)

Drying to
interior
and exterior

Figure IV.11
Drying to Interior and Exterior

- If wood siding is used in this assembly with the damp spray cellulose, furring strips should be used to provide an airspace to promote drying and the wood siding should be back-primed to prevent wetting from the back side.
- The airspace associated with the back of vinyl or aluminum siding, due to its profile, permits drying of the wall assembly.
- Polyethylene on the inside of building assemblies in cold, mixed-humid, mixed-dry, hot-humid and hot-dry climates is not generally a good idea.

12 IV

Appendices

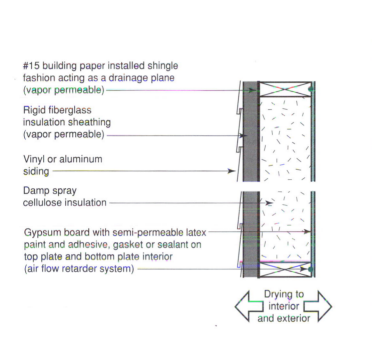

#15 building paper installed shingle fashion acting as a drainage plane (vapor permeable)

Rigid fiberglass insulation sheathing (vapor permeable)

Vinyl or aluminum siding

Damp spray cellulose insulation

Gypsum board with semi-permeable latex paint and adhesive, gasket or sealant on top plate and bottom plate interior (air flow retarder system)

Drying to interior and exterior

Figure IV.12
Drying to Interior and Exterior

- If wood siding is used in this assembly with the damp spray cellulose, furring strips should be used to provide an airspace to promote drying and the wood siding should be back-primed to prevent wetting from the back side.
- The airspace associated with the back of vinyl or aluminum siding, due to its profile, permits drying of the wall assembly.
- Polyethylene on the inside of building assemblies in cold, mixed-humid, mixed-dry, hot-humid and hot-dry climates is not generally a good idea.

IV 12

Appendices

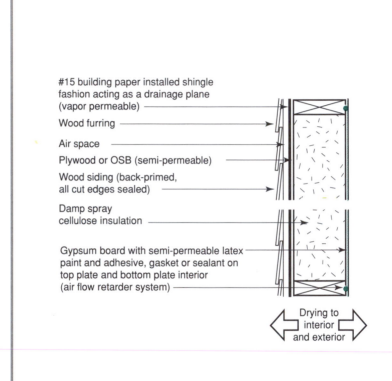

#15 building paper installed shingle fashion acting as a drainage plane (vapor permeable)

Wood furring

Air space

Plywood or OSB (semi-permeable)

Wood siding (back-primed, all cut edges sealed)

Damp spray cellulose insulation

Gypsum board with semi-permeable latex paint and adhesive, gasket or sealant on top plate and bottom plate interior (air flow retarder system)

Drying to interior and exterior

Figure IV.13
Drying to Interior and Exterior
- If vinyl or aluminum siding is used in this assembly, wood furring providing an airspace is not necessary.
- Polyethylene on the inside of building assemblies in cold, mixed-humid, mixed-dry, hot-humid and hot-dry climates is not generally a good idea.

12 IV

Appendices

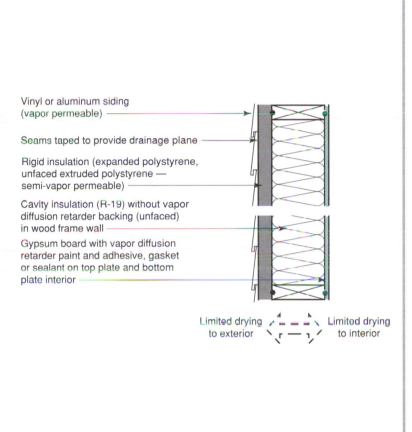

Vinyl or aluminum siding (vapor permeable)

Seams taped to provide drainage plane

Rigid insulation (expanded polystyrene, unfaced extruded polystyrene — semi-vapor permeable)

Cavity insulation (R-19) without vapor diffusion retarder backing (unfaced) in wood frame wall

Gypsum board with vapor diffusion retarder paint and adhesive, gasket or sealant on top plate and bottom plate interior

Limited drying to exterior — — — Limited drying to interior

Figure IV.14
Limited Drying to Exterior and Interior

- Although paint is used as an interior vapor diffusion retarder (1 to 2 perms) it is not as impermeable as a polyethylene vapor diffusion retarder (0.3 to 0.5 perms) so that some drying to the interior is possible.
- The semi-permeable rigid insulations permit some drying to the exterior.
- If wood siding is used, it should be installed over furring strips and be back-primed, all cut edges sealed.

IV 12

Appendices

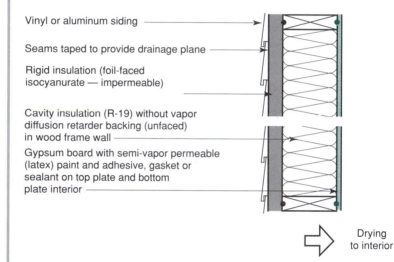

Vinyl or aluminum siding

Seams taped to provide drainage plane

Rigid insulation (foil-faced isocyanurate — impermeable)

Cavity insulation (R-19) without vapor diffusion retarder backing (unfaced) in wood frame wall

Gypsum board with semi-vapor permeable (latex) paint and adhesive, gasket or sealant on top plate and bottom plate interior

Drying to interior

Figure IV.15
Drying to Interior
- The semi-vapor permeable latex paint permits drying to the interior
- If wood siding is used, it should be installed over furring strips and be back-primed, all cut edges sealed.

12 IV

Appendices

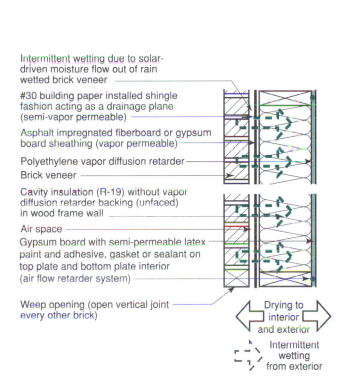

Intermittent wetting due to solar-driven moisture flow out of rain wetted brick veneer

#30 building paper installed shingle fashion acting as a drainage plane (semi-vapor permeable)

Asphalt impregnated fiberboard or gypsum board sheathing (vapor permeable)

Polyethylene vapor diffusion retarder

Brick veneer

Cavity insulation (R-19) without vapor diffusion retarder backing (unfaced) in wood frame wall

Air space

Gypsum board with semi-permeable latex paint and adhesive, gasket or sealant on top plate and bottom plate interior (air flow retarder system)

Weep opening (open vertical joint every other brick)

Drying to interior and exterior

Intermittent wetting from exterior

Figure IV.16
Drying to Interior and Exterior

- Polyethylene on the inside of building assemblies in cold, mixed-humid, mixed-dry, hot-humid and hot-dry climates is not generally a good idea.
- A rigid, impermeable or semi-permeable insulating sheathing can be used to prevent the wall cavity from getting wet due to solar-driven moisture allowing the removal of the interior polyethylene vapor diffusion retarder
- The heavy #30 building paper (semi-vapor permeable) is preferred over permeable building papers when used with wood siding, brick or stucco due to the potential for moisture flow through the permeable building papers under solar heating with rain wetted claddings.

IV 12

Appendices

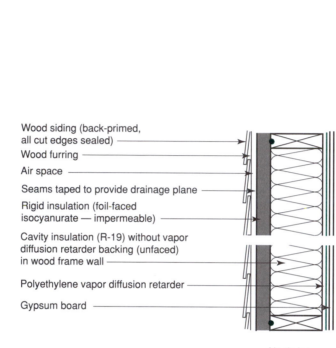

Wood siding (back-primed, all cut edges sealed)

Wood furring

Air space

Seams taped to provide drainage plane

Rigid insulation (foil-faced isocyanurate — impermeable)

Cavity insulation (R-19) without vapor diffusion retarder backing (unfaced) in wood frame wall

Polyethylene vapor diffusion retarder

Gypsum board

No drying

Figure IV.17
No Drying

- Only dry materials should be used in the construction of this wall assembly
- If vinyl or aluminum siding is used in this assembly wood furring providing an airspace is not necessary
- Airspace in this assembly behind the wood siding is to permit drying of the wood siding
- This is an extremely unforgiving wall assembly
- In a mixed-humid climate, this wall should be constructed without the polyethylene vapor diffusion retarder to permit drying to the interior

12 IV

Appendices

Appendix V

Windows

Benefits of High Performance Windows in Mixed Climates

Heating and Cooling Season Savings: In winter, this means that windows are no longer a major source of heat loss. In summer, this means that window heat gain can be managed as long as the right kind of glazing and shading techniques are in place.

Comfort: In winter, high performance windows will result in a higher interior window temperatures and fewer uncomfortable drafts. In summer, high performance windows reduce the direct solar radiation.

Reduced Condensation: During winter, high performance windows create warmer interior glass surfaces, reducing frost and condensation.

Improved Daylight and View: Windows with spectrally selective low-E coatings and tints can provide better solar heat gain reduction than bronze- or gray-tinted glass, with a minimal loss of visible light. This means that views can be clearer and less obstructed by shades.

Lower Mechanical Equipment Costs: Using windows that significantly reduce solar heat gain means that mechanical equipment costs for the air conditioning system can be reduced as well. This also means that utility companies benefit because peak loads are reduced.

The material in this appendix is drawn from the book, "Residential Windows: New Technologies and Energy Performance," by John Carmody, Stephen Selkowitz, and Lisa Heschong (W.W. Norton, 1996). This effort was supported by the U.S. Department of Energy.

Selecting an Energy-Efficient Window

In mixed climates where houses require both heating and cooling, conventional windows have represented a major source of unwanted heat loss in winter, and a major source of unwanted heat gain in summer. In recent years, low-E coatings and other improvements have revolutionized window performance. This appendix is intended to help homeowners select energy efficient windows in a new home or replace windows in an existing home. The term *window* is used generically in this publication and also refers to glass doors and skylights.

There are three tools to assist you in selecting energy-efficient windows (explained on the following pages). Which tools you use depends on the amount of time and energy you want to invest in the decision.

ENERGY EFFICIENT WINDOW IMPROVEMENTS

- Low-E coatings reduce winter heat loss

- Spectrally selective low-E coatings and tints reduce summer heat gain without losing too much light & view

- Low-conductance gas fills reduce winter heat loss

- Low-conductance spacers reduce winter heat loss

- Thermally improved frames reduce winter heat loss

- Improved weatherstripping reduces air leakage

12 V

Appendices

Figure V.1
Technological Advances in Windows

1. Look for the **Energy Star** to identify efficient products. This takes the least amount of time and ensures an energy efficient product.

2. Look for the **NFRC Label**. Based on the label, compare window properties such as U-factor and Solar Heat Gain Coefficient to find the most efficient products among those with Energy Stars. This takes a little more effort but can be relatively easy using the guidelines on the next page.

3. If you want a reasonably accurate estimate of energy savings based on your specific house, climate, and utility costs, use a computer program such as **RESFEN.** This is by far the best way to make a selection but will take a little more time and the use of a computer. Using and obtaining **RESFEN** is explained on the following pages.

In a new or existing home, selecting the right window depends on many other design issues. The orientation and shading of a window affect its performance, for example.

Look for the Energy Star

The Department of Energy (DOE) and the Environmental Protection Agency (EPA) have developed an Energy Star designation for products meeting certain energy performance criteria. Since energy efficient performance of windows, doors, and skylights varies by climate, product recommendations are given for the U.S. climate zones similar to those shown on the map in the front of this book. For making comparisons among Energy Star products, use the NFRC label or directory.

Look for the NFRC Label

The National Fenestration Rating Council (NFRC) has developed a window energy rating system based on whole product performance. All Energy Star products must have an NFRC label. The NFRC label provides the only reliable way to determine the basic energy-related properties of the window unit and to compare products (see guidelines below).

NFRC labels on window units give ratings for U-factor, solar heat gain coefficient (SHGC), visible light transmittance (VT), air infiltration rates (AL) and an annual heating and cooling rating (HR and CR).

Check local building codes for minimum window property requirements in your location. Most energy codes have limits for maximum U-factor allowed. Some also have requirements for SHGC and air leakage. Then use the following guidelines to compare products.

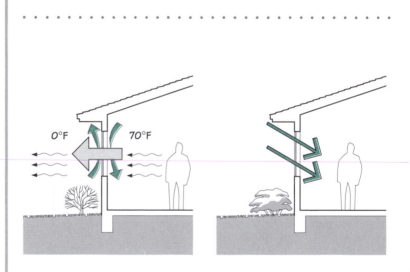

Figure V.2
Heat Flow

- Indicated by U-factor (U-value).

- Select windows with a U-factor of 0.40 or less. The larger your heating bill, the more important a low U-factor becomes. Some double-glazed low-e products have U-factors below 0.30.

Figure V.3
Solar Heat Gain

- Indicated by solar heat gain coefficient (SHGC).

- In mixed climates, select windows with a SHGC of 0.55 or less. If you have significant air conditioning costs or summer overheating problems, look for SHGC values of 0.40 or less.

12 **V**

Appendices

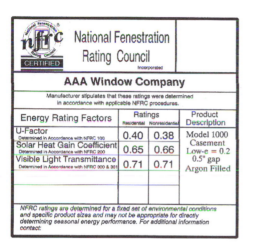

National Fenestration Rating Council Incorporated		
AAA Window Company		
Manufacturer stipulates that these ratings were determined in accordance with applicable NFRC procedures.		
Energy Rating Factors	Ratings Residential / Nonresidential	Product Description
U-Factor Determined in Accordance with NFRC 100	0.40 / 0.38	Model 1000
Solar Heat Gain Coefficient Determined in Accordance with NFRC 200	0.65 / 0.66	Casement Low-e = 0.2
Visible Light Transmittance Determined in Accordance with NFRC 300 & 301	0.71 / 0.71	0.5" gap Argon Filled

NFRC ratings are determined for a fixed set of environmental conditions and specific product sizes and may not be appropriate for directly determining seasonal energy performance. For additional information contact:

Figure V.4
NFRC Label

Figure V.5
Infiltration

- Indicated by air leakage rating (AL).

- The air leakage rating (AL) is an important window property in cold climates. Select a window with an AL of 0.30 or below (units are cfm/ sq ft).

Figure V.6
Daylight

- Indicated by visible light transmittance (VT).

- A window with VT*glass* above 0.70 (for the glass only) is desirable to maximize daylight and view. This translates into a VT*window* above 0.50 (for the total window including frame).

V **12**

Appendices

Comparing Annual Energy Performance

Using a computer program to compare window options is the only method of obtaining reasonable estimates of the heating and cooling costs for your climate, house design, and utility rates. One computer program available for this purpose is RESFEN (see end of appendix). The results from a computer simulation are shown for four possible window choices for a typical house in St. Louis, Missouri.

Window A is a typical clear, single-glazed unit and Window B is a typical clear, double-glazed unit. Window C has a low-E coating, while Window D has a spectrally selective low-E coating. Window C is designed to reduce winter heat loss (low U-factor) and provide winter solar heat gain (high SHGC). Window D also reduces winter heat loss (low U-factor) but it reduces solar heat gain as well (low SHGC).

To determine a payback for investing in energy efficient windows, it is necessary to factor in not only annual energy savings but possible savings from reducing the size of the mechanical equipment (in new construction). Then consider other benefits such as greater comfort.

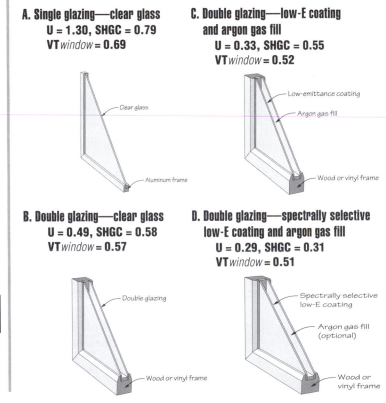

A. Single glazing—clear glass
U = 1.30, SHGC = 0.79
VT *window* = 0.69

Clear glass

Aluminum frame

C. Double glazing—low-E coating and argon gas fill
U = 0.33, SHGC = 0.55
VT *window* = 0.52

Low-emittance coating

Argon gas fill

Wood or vinyl frame

B. Double glazing—clear glass
U = 0.49, SHGC = 0.58
VT *window* = 0.57

Double glazing

Wood or vinyl frame

D. Double glazing—spectrally selective low-E coating and argon gas fill
U = 0.29, SHGC = 0.31
VT *window* = 0.51

Spectrally selective low-E coating

Argon gas fill (optional)

Wood or vinyl frame

12 V

Appendices

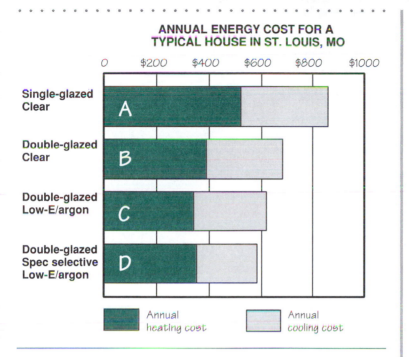

ANNUAL ENERGY COST FOR A TYPICAL HOUSE IN ST. LOUIS, MO

Single-glazed Clear — A

Double-glazed Clear — B

Double-glazed Low-E/argon — C

Double-glazed Spec selective Low-E/argon — D

Annual heating cost

Annual cooling cost

Figure V.7
Energy Cost Comparison with Different Windows

Note: The windows used in this comparison are illustrative of typical products, but there are many other windows with excellent heating and cooling season performance not shown here. The annual energy performance figures shown here were generated using the DOE 2.1E computer program for a typical 1,540 sq. ft. house with 231 sq. ft. of window area (15% of floor area). The windows are equally distributed on all four sides of the house and are unshaded. U-factor, SHGC, and VT are for the total window including frames.

In applying these typical results to your particular situation, remember that our example is a relatively small house (1,500 sq. ft.) with an average amount of window area (231 sq. ft.). The fuel and electricity rates used to calculate energy costs are national averages (Cooling season: $0.095/kWh for electricity, Heating season: $6.40/Mbtu for natural gas). St. Louis, Missouri is only one example of a mixed climate with 4,948 HDD (heating degree days) and 17,843 CDH (cooling degree hours). Instead of drawing conclusions from average conditions such as these, the best way to compare different windows is by using a computer tool such as RESFEN where you can base decisions on your own house design, climate, and fuel costs for your area.

V 12

Appendices

Window Description	Total Window Unit			Center of Glass Only		
	U-value	SHGC	AL	U-value	SHGC	VT
Single-glazed Clear glass Aluminum frame*	1.30	0.79	0.98	1.11	0.86	0.90
Double-glazed Clear glass Aluminum frame**	0.64	0.65	0.56	0.49	0.76	0.81
Double-glazed Bronze tinted glass Aluminum frame**	0.64	0.55	0.56	0.49	0.62	0.61
Double-glazed Clear glass Wood or vinyl frame	0.49	0.58	0.56	0.49	0.76	0.81
Double-glazed Low-E (high solar gain) Argon gas fill Wood or vinyl frame	0.33	0.55	0.15	0.30	0.74	0.74
Double-glazed Low-E (medium solar gain) Argon gas fill Wood or vinyl frame	0.30	0.44	0.15	0.26	0.58	0.78
Double-glazed Spectrally selective coating (low solar gain) Argon gas fill Wood or vinyl frame	0.29	0.31	0.15	0.24	0.41	0.72
Triple-glazed Clear glass Wood or vinyl frame	0.34	0.52	0.15	0.31	0.69	0.75
Triple-glazed Two low-E coatings Krypton gas fill Wood or vinyl frame	0.15	0.37	0.08	0.11	0.49	0.68

* No thermal break in frame.
** Thermal break in frame.
All values for total windows are based on a 2-foot by 4-foot casement window.
Units for all U-values are Btu/hr-sq ft-°F.
SHGC = solar heat gain coefficient.
AL = air leakage in cubic feet per minute per square foot of unit.
VT = visible transmittance.

Appendices

Figure V.8
Properties of Some Typical Window Types

Designing Your Home for Energy Efficiency

In a mixed climate, the energy-efficiency goals must address both the heating and cooling season needs. In winter, the goals are to decrease winter heat loss through the building envelope and to provide useful solar heat gain. In summer, the most important energy-efficiency goal is to decrease summer heat gain through the building envelope. A number of strategies to accomplish these goals are listed below. The publications and resources listed at right provide more guidance on these issues.

Guidelines for Decreasing Winter Heat Loss

The strategies to decrease winter heat loss differ depending on whether you have traditional clear glazing or high performance windows.

- With clear single- or double-glazed windows, avoid extensive window areas to reduce winter heat loss. When using high performance products, however, window area can be increased without a significant heating energy penalty.

- With clear single- or double-glazed windows, use thermal shades or movable insulation over windows to reduce nighttime heat loss. Thermal shades over high performance windows are not necessary to improve energy performance.

Guidelines for Providing Winter Heat Gain

- Face windows to the south to provide maximum solar gain in winter. However, when using high-performance (low U-factor) windows, orientation has less impact on heating energy performance in a typical house without significant thermal mass.

Guidelines for Decreasing Summer Heat Gain

- Design the layout so that windows and living areas do not face the hot western sun.

- Minimize window area on west and southwest orientations to reduce summer solar heat gain.

- Use overhangs or other architectural and landscape elements to prevent sunlight from reaching windows.

- Use drapes, blinds, shades, or other interior treatments to reduce heat gain through windows.

Appendices

Publications

Carmody, J., S. Selkowitz, and L. Heschong. *Residential Windows: New Technologies and Energy Performance.* New York, NY: W.W. Norton & Company, 1996.

Resources

Efficient Windows Collaborative.
www.efficientwindows.org

For more information on the Efficient Windows Collaborative, contact:

Alliance to Save Energy
1200 18th Street, N.W., Suite 900
Washington, D.C. 20036
phone: 202-857-0666, fax: 202-331-9588
www.ase.org

Energy Star Program
www.energystar.gov

The Home Energy Saver
An interactive web site with housing information.
http://eande.lbl.gov/CBS/VH/vh.html

National Fenestration Rating Council (NFRC)
1300 Spring Street, Suite 500
Silver Spring, MD 20910
Phone: (301) 589-NFRC
www.nfrc.org

U.S. Department of Energy:
1-800-DOE-EREC
www.eren.doe.gov

RESFEN—A computer program for calculating the annual cooling and heating energy use and costs due to window selection. RESFEN is available from Lawrence Berkeley National Laboratory. Fax a request to Resfen Request at 510-486-4089 or e-mail your request to plross@lbl.gov.

12 V

Appendices

Appendix VI

Air Leakage Testing, Pressure Balancing and Combustion Safety

Air leakage testing is a method for determining the total leakage area of a building envelope or the leakage of air distribution systems (duct-work leakage in ducted forced air space conditioning systems). Air leakage testing is not a method for determining the actual air leakage or air change which occurs through the building envelope under the influence of air pressure differences created by wind, stack action (the buoyancy of heated air) and mechanical systems (duct leakage and un-balanced forced air systems).

Air leakage testing for both building envelopes and ductwork is based on the fundamental properties of air flow through openings. The amount of air flow through an opening is determined by two principle factors:

- the area/size/geometry of the opening; and

- the air pressure difference across the opening.

The three parameters — air flow, area and air pressure difference — can be related to each other by applying a simple mathematical relationship. Measuring two of the three parameters and applying the mathematical relationship can determine the third parameter. For example, if the air flow through an opening is measured, as well as the air pressure difference across the opening when air flow is occurring, the area of the opening can be calculated by applying the mathematical relationship. Applying this relationship to a flow rate of 1,000 cfm through an opening with an air pressure differential of 50 Pascals across the opening obtains a mathematically calculated area of approximately 1-square-foot. In other words 1,000 cfm air flow occurs through a 1-square-foot opening as a result of an air pressure difference of 50 Pascals.

Air Leakage Testing of Building Envelopes

Air leakage testing of building envelopes involves placing a large calibrated fan in an exterior door and creating an air flow through the fan. The calibrated fan is often referred to as a "blower door". Exhausting air depressurizes the building. Supplying air pressurizes the building. When air is exhausted from a building through a blower door, air leaks into the building through openings to replace the air exhausted. If sufficient air is exhausted to overcome any naturally occurring pressures, the quantity of air exhausted will equal the quantity of air supplied. The quantity of air exhausted through the blower door can be readily measured. This exhaust quantity can be equated to the air leaking into the building envelope through all of the openings in the building envelope. If the air pressure difference between the interior of the building envelope and the exterior is also measured, this can be used to approximate the air pressure difference across all of the openings in the building envelope.

By determining the quantity of air exhausted and the air pressure difference across the building envelope, the combined air flow through all of the openings in the building envelope as well as the air pressure difference across all of the openings is known. Applying a mathematical relationship converts the combined air flow and air pressure difference to the combined leakage area of all of the openings in the building envelope. In this manner a blower door can determine the combined area of all of the openings in a building envelope, including random cracks, flaws, openings built into the building as a result of the building process, without actually determining where the leaks and openings are.

The procedure can also be applied to building envelopes pressurized by calibrated fans. Either depressurization or pressurization can be employed. Depressurization is more common in building envelope leakage testing as a result of tradition rather than accuracy (the procedures were initially popularized in cold climates where pressurization during the heating season typically created comfort problems during testing).

Air leakage test results are expressed several ways. Test results can be presented as a flow rate at 50 Pascals air pressure difference (CFM50). In other words, the volume flow rate of air extracted out of the building envelope necessary to depressurize the building envelope 50 Pascals relative to the exterior is measured and reported. The combined or equivalent leakage area of the building (EqLA) in square inches can be determined from the CFM50 measurement by the application of the mathematical relationship.

The combined leakage area of a building envelope can also be compared to the total surface area of the building envelope using a parameter called a leakage ratio. Leakage ratios are typically expressed as square inches of leakage for every 100 square feet of building envelope area (or $cm^2/100m^2$ SI). In this manner the measured EqLA is related to the measured surface area of the building envelope.

Test results can also be expressed in the form of air changes per hour at a pressure difference of 50 Pascals (ach @ 50 Pa). In this approach, the volume flow rate extracted out of the building is related to the volume of the building envelope. For example, consider a blower door extracting air out of a building envelope at a rate of 1,000 cubic feet per minute establishing an air pressure difference of 50 Pa. This is equivalent to an air extraction rate of 60,000 cubic feet per hour. If the volume of the conditioned space is 10,000 cubic feet, this would result in 6 air changes per hour at a 50 Pascal air pressure difference (60,000/ 10,000 = 6). A flow rate of 1,000 cfm through an opening with a 50 Pascals air pressure difference was previously determined to require a 1 foot square opening. In other words, for this particular volume of building, 6 ach @ 50 Pa is equivalent to 1 foot square of leakage or an EqLA of 1 foot square or 144 square inches. If the surface area of the building envelope was approximately 3000 square feet, the leakage ratio would be approximately 4.8 square inches of leakage for every 100 square feet of building envelope surface area (3000/100 = 30 and 144/ 30 = 4.8).

The following information all describes the same building. It came from the blower door test conducted on the building described in the previous example.

$$CFM50 = 1000$$

$$EqLA = 144 \text{ in}^2$$

6 ach @ 50 Pa (where building volume is 10,000 ft^3)

Leakage Ratio = 4.8 in^2/100 ft^2 (where building envelope surface area is 3000 ft^2)

Not all of these values are always determined or recorded. All can be related to each other and the building envelope tested. The leakage ratio value is the most descriptive as it can be used to compare buildings of differing volumes and surface areas to each other. Since the other values are related to specific buildings with given building volumes and building surface areas, comparisons between buildings of diverse construction are less meaningful. However, where buildings are of approximately the same floor area and volume, all of the values provide reasonably comparative information.

Air leakage testing of building envelopes is typically conducted as a method of quality control to ensure that control of air flow occurs in constructed buildings. It is also used to identify leakage areas which may have been missed during construction thereby facilitating repairs and remediation.

Air Leakage Testing of Air Distribution Systems

Air leakage testing of air distribution systems is similar to air leakage testing of building envelopes in that both procedures involve using a calibrated fan to create an air pressure difference. In addition, both procedures require that air flows through the calibrated fans and pressure differentials be determined.

Air leakage testing of air distributions systems involves sealing the supply and return registers and depressurizing the system using a calibrated fan. In this approach, the ductwork and the air handler is considered a closed system. The calibrated fan, sometimes referred to as a "duct blaster" is typically attached to the air handler. The quantity of air moved by the duct blaster to pressurize or depressurize the system is directly related to the leakage area of the system. By determining the quantity of air supplied by the calibrated fan and the air pressure difference between the duct work and the conditioned space, the combined air flow through all of the leakage openings in the air distribution system is determined.

Air leakage test results for air distribution systems are often presented as the flow rate through the calibrated fan required to pressurize or depressurize the duct system to a specific pressure differential. For example CFM25 values are typical. A reading of CFM25 = 60 cfm translates to: an air flow rate of 60 cfm through the calibrated fan (duct blaster) depressurized the air distribution system (ductwork and air handler) to a negative of 25 Pascals (0.1" w.c.) relative to the conditioned space.

Air leakage testing of air distribution systems is typically conducted as a method of quality control to ensure control of air flow occurs in air distribution systems. It is also used to identify leakage areas thereby facilitating repairs and remediation.

Pressure Balancing

Air pressure differentials between conditioned spaces and the surroundings, as well as between rooms and between building assemblies and rooms, impact the health, safety and durability of the building envelope.

Infiltration of humid air during cooling periods is a concern as is the exfiltration of interior heated, moist air during heating periods. Infiltration of soil gas (moisture, radon, pesticides, other) below grade or from crawlspaces and slabs is a concern throughout the year. High interior negative air pressures can lead to the spillage and backdrafting of combustion appliances such as fireplaces, wood stoves, combustion water heaters and furnaces. Finally, significant depressurization can lead to flame roll out and fire in some combustion appliances such as furnaces and water heaters.

With respect to combustion appliances, installation of appliances which are not sensitive to negative air pressures as well as limiting the negative air pressures which can occur within building enclosures is an appropriate strategy for control.

With respect to the infiltration of exterior pollutants such as soil gas, pesticides, radon and below grade moisture, limiting the negative air pressure which can occur, providing positive air pressurization of building enclosures (or portions of building enclosures) as well as providing subslab or crawlspace depressurization are appropriate strategies for control.

With respect to the exfiltration of interior moisture during heating periods, constructing building assemblies which are forgiving and/or tolerant of interior moisture, as well as limiting the positive air pressure which can occur during heating periods (and/or providing negative pressures during heating periods) are appropriate strategies for control.

The maximum pressurization or depressurization limit relative to exterior (ambient) conditions can vary based on the climate, time of year (season) and combustion appliances installed. In general, pressure differences induced by mechanical equipment should be limited to 5 Pascals.

In all cases the maximum interzonal pressurization or depressurization should be limited to less than 3 Pascals.

Ducted forced air distribution systems are traditionally viewed as interior circulation systems which move air from place to place within a conditioned space, with a neutral effect on the pressure differences between the interior and exterior. However, as a result of installation

VI 12

Appendices

practices and design/sizing faults ducted forced air distribution systems can have significant effects on air pressure relationships.

Duct leakage can result in either pressurization or depressurization of entire conditioned spaces or specific rooms. Duct leakage can significantly increase space conditioning energy requirements. Incorrect duct sizing, distribution layout or lack of adequate returns can lead to pressurization and depressurization of rooms and interstitial spaces. These effects should be limited to less than 3 Pascals positive or negative relative to the exterior or between rooms and/or interstitial (between two surfaces) cavities within building enclosures.

Exhaust fans and appliances such as whole house fans, attic ventilation fans, indoor grills, clothes dryers, kitchen exhaust range hoods can also significantly alter air pressure relationships. These effects should also be limited to less than 3 Pascals positive or negative relative to the exterior or between rooms and/or interstitial cavities within building enclosures (except in the case of whole house fans and indoor grills). Where whole house fans and/or indoor grills are operating, no other combustion appliances or heating and/or mechanical cooling systems should be in simultaneous use. Furthermore, windows and/or doors should be opened when whole house fans and/or indoor grills are operating.

Testing Pressure Differentials and Commissioning

Air pressure relationships between conditioned spaces and the exterior, as well as between rooms and between rooms and interstitial spaces should be measured under all operating conditions. Equipment should be cycled on and off at all speed settings. Interior doors should be both opened and closed during all testing. Measurements are typically taken with a digital micromanometer. Where any air pressure differential greater than 3 Pascals is measured, remediation work and/or adjustments to equipment, or the building envelope will be necessary.

Combustion Safety

If combustion appliances are selected, they should not interact aerodynamically with the building. In other words, changing interior air pressure differentials should not be able to influence the operation of combustion appliances. In order to meet this requirement only sealed combustion, power vented, induced draft or direct vented combustion appliances should be used for space conditioning and domestic hot water.

Gas cook tops and ovens should be only installed in conjunction with direct vented (to the exterior) exhaust range hoods. Recirculating range hoods should be avoided even in the absence of combustion appliances as they become breeding grounds for biological growth and a source of odors.

Fireplaces and wood stoves should only be installed with their own correctly sized air supply from the exterior. Fireplaces should also be provided with tight-fitting glass doors.

High interior negative air pressures can lead to the spillage and backdrafting of combustion appliances and significant depressurization can lead to flame roll-out and fire. In all enclosures containing combustion appliances, the maximum depressurization relative to the exterior should be limited to less than 3 Pascals.

All combustion appliances (except fireplaces and wood stoves) should be tested for carbon monoxide production prior to occupancy and on a yearly basis thereafter. Carbon monoxide production of any appliance should not exceed 50 ppm. All measurements should be taken in the draft hood or vent before exhaust gases are mixed with dilution air. The installation of household carbon monoxide detectors is recommended.

VI 12

Appendices

Appendices

Appendix VII
Goals, Objectives and Criteria for Energy and Resource Efficient Buildings

Introduction

The Energy Efficient Building Association (EEBA) has developed these goals, objectives and criteria for energy and resource efficient buildings. They provide guidance for design, construction and comprehensive rehabilitation (gut-rehab) of low-rise residential and small commercial buildings less than 20,000 square feet (1900 m²) floor area.

Goals

Energy Efficiency

To promote building practices that result in a substantial reduction in energy use for space conditioning, water heating, lighting and appliance operation.

> *Improved energy efficiency can reduce the environmental impact of the built environment, improve economic well being and promote global stability.*

Occupant Safety

To promote building practices that result in an improvement in fire and structural safety.

Improved construction practices can reduce the risk from earthquakes, hurricanes, floods and fires.

Occupant Health

To promote building practices that result in an improvement in the indoor environment.

Improved construction practices can reduce the risk from building related illness and sick building syndrome.

Durability

To promote building practices that prolong the useful service life of buildings, reduce maintenance and promote serviceability.

Rehabilitation and replacement of damaged components and structures results in the inefficient use of resources. Improper moisture control can lead to premature failure of building components and can contribute to poor environmental conditions for occupants.

Occupant Comfort

To promote building practices that improve thermal comfort, daylighting, lighting, humidity control, odor control, noise control and vibration control.

Providing comfort for building occupants is one of the fundamental requirements of shelter.

Environmental Impact

To promote building practices that reduce the impact on the local and global environment.

The impacts of the built environment on the planetary environment make it necessary to make informed, environmentally responsible choices during the construction process.

Objectives

Energy Efficiency

Energy efficient and resource efficient construction should address the following objectives for design, construction, commissioning, operation and maintenance.

A. Building Structure

Thermal transmission through heat loss and heat gains should be reduced by the specification and installation, with proper attention to detail and quality assurance, of increased levels of thermal insulation. Insulation systems should be installed such that they reduce convective, conductive and radiative heat losses and gains. Thermal anomalies such as thermal bridges should be minimized.

Moisture gain resulting in decreased thermal and structural performance should be controlled. Air flow retarder systems and vapor diffusion retarders should be used to protect the building envelope from uncontrolled air and moisture flow.

Thermal transmission through convective heat loss and gain driven by "wind-washing" should be reduced by the specification and installation, with proper attention to detail and quality assurance, of an external air barrier system or external "weather barrier".

Fenestration systems should be selected according to climate, building orientation, interior comfort, daylighting, ventilation, furnishing durability and egress requirements.

B. Mechanical Systems

Indoor air quality should be facilitated by the installation of a controlled mechanical ventilation system. Heat recovery is recommended in severe heating climate zones.

Only sealed combustion or power vented direct combustion appliances should be installed in occupied spaces. Gas cooktops and gas ovens should only be installed in conjunction with exhaust fans.

Thermal and peak load reductions derived from improving levels of insulation, airtightness and fenestration performance of the building envelope should be evaluated in the sizing of equipment.

The domestic hot water system should meet high efficiency standards. Options for reducing water consumption are recommended. Solar energy for hot water heating should be considered.

Efficient illumination design and lighting systems should be used. Natural lighting of spaces should be considered prior to specifying electric illumination systems. Lighting designs and controls should consider the availability of natural light. Occupancy sensors should be considered for foyers, utility room, basements, garages and other spaces. Hard wired general area lighting should employ fluorescent fixtures. Other lighting fixtures should use compact fluorescent lamps.

Major appliances should meet high energy efficiency standards using current appliance ratings.

C. Occupant Considerations

A comprehensive operations manual should be provided to occupants which includes necessary operating, maintenance and repair information so that the performance of the building can be maximized.

Occupant Safety

In no case should the application of energy efficient or resource efficient design or construction strategies, materials, equipment or appliances violate safety codes and standards.

A. Building Structure

Recognized structural design shall be employed to resist live, static and wind loads.

B. Mechanical Systems

Mechanical systems shall be designed and constructed to facilitate occupant safety.

C. Occupant Considerations

Information relating to the safe operation of the building and mechanical systems shall be provided to occupants. Information relating to safe maintenance of installed mechanical systems shall also be provided.

Occupant Health

Energy efficient and resource efficient construction should provide a healthy living and working environment.

A. Building Structure

Selection of construction materials that have low emission rates of toxic materials; foundations designed to exclude entry of soil gas; and implementation of moisture control measures are recommended.

B. Mechanical Systems

A controlled mechanical ventilation system should be provided to facilitate occupant health.

C. Occupant Considerations

Information relating to the healthy operation of the building and its mechanical systems should be provided to the occupants.

Durability

Energy efficient and resource efficient construction should include the following moisture control measures in order to provide long term performance and durability.

A. Building Structure

The building envelope should provide mechanisms to control the migration of moisture in the liquid and vapor form.

Building materials and components should be protected from rain, snow and other moisture sources during storage on site, construction and commissioning of the building.

B. Mechanical Systems

Controlled ventilation, mechanical cooling or dehumidification systems should be provided to maintain acceptable indoor relative humidity. Such systems and their controls should maintain humidity in the range of 25 to 60 percent. Source control of moisture should be used where possible.

C. Occupant Considerations

Instructions for the proper use and maintenance of moisture control systems should be provided to occupants.

Occupant Comfort

Energy efficient and resource efficient construction should provide a comfortable living and working environment.

A. Building Structure

The building envelope should facilitate the comfort of occupants.

B. Mechanical Systems

The mechanical systems should facilitate the comfort of occupants.

C. Occupant Considerations

Information relating to the comfortable operation of the building and its mechanical systems should be provided to the occupants.

Environmental Impact

Energy efficient and resource efficient construction should minimize the impact on the environment. Design and construction strategies which account for full life-cycle energy consumption and resource utilization—including the reuse, recycling and reconfiguration of materials and practices—should be used.

A. Building Structure

The building envelope should be deployed on its site and in its local environment in an environmentally sensitive fashion.

Use of virgin materials or materials with low recycled content should be minimized.

On-site reuse of surplus construction materials should be provided. Recycling of materials should be maximized.

B. Mechanical Systems

The energy efficiency of mechanical conditioning systems should be maximized.

C. Occupant Considerations

Information relating to the resource efficient operation and performance of the building should be provided to the occupants. Measures facilitating the recycling of consumer waste should be utilized.

Criteria

The following criteria are recommended for the design and construction of energy and resource efficient buildings.

Component Criteria

A. Building Structure

Overall energy consumption for heating, cooling and water heating should meet Energy Star requirements (30% improvement over a standard reference home based on the envelope and equipment requirements of the 1993 Model Energy Code) as determined by an accredited home energy rating system procedure.

Air leakage of buildings (determined by pressurization testing) should be less than 2.5 square inches/100 square feet leakage ratio (CGSB, calculated at a 10 Pa pressure differential); or, 1.25 square inches/100 square feet leakage ratio (ASTM, calculated at a 4 Pa pressure differential); or, 0.25 cfm/square foot of building envelope surface area @ 50 Pa.

B. Mechanical Systems

Controlled mechanical ventilation at a minimum base rate of 20 cfm per master bedroom and 10 cfm for each additional bedroom will be provided when the building is occupied.

A capability to increase the base rate ventilation on an intermittent basis to 0.05 cfm per square foot of conditioned areas will also be provided.

Intermittent spot ventilation of 100 cfm will be provided for each kitchen. Intermittent spot ventilation of 50 cfm or continuous ventilation of 20 cfm will be provided for each washroom/bathroom.

Positive indication of shut-down or improper system operation for the base rate ventilation will be provided to occupants.

Mechanical ventilation shall use less than 0.5 watt/cfm for ventilation systems without heat recovery or less than 1.0 watt/cfm for ventilation systems with heat recovery.

Mechanical ventilation system airflow should be tested during commissioning of the building.

Heat recovery on controlled mechanical ventilation is recommended in severe heating climate zones. Heat recovery rates of heat recovery ventilators should be greater than 65 percent, including effectiveness of distribution.

VII 12

Appendices

Total ductwork leakage for ducts distributing conditioned air should be limited to 10.0 percent of the total air handling system rated air flow at high speed determined by pressurization testing at 25 Pa.

Ductwork leakage to the exterior for ducts distributing conditioned air should be limited to 5.0 percent of the total air handling system rated air flow at high speed determined by pressurization testing at 25 Pa.

Only sealed combustion or power vented direct combustion appliances should be installed in occupied spaces. These appliances must be rated to vent properly at largest expected negative pressure. Gas cooktops and gas ovens should only be installed in conjunction with exhaust fans.

Major appliances should meet high energy efficiency standards using current appliance ratings. Select only those appliances in the top one-third of the DOE Energy Guide rating scale.

Lighting power density should not exceed 1.0 Watts per square foot.

C. Occupant Considerations

Systems that provide control over space conditioning, hot water or lighting energy use should be clearly marked. Information relating to the operation and maintenance of such systems should be provided to occupants.

The designer and general contractor should provide comprehensive information to occupants relating to the safe, healthy, comfortable operation of the building and mechanical systems.

Indoor Environment Criteria

Energy efficient and resource efficient construction should provide comfortable indoor conditions as defined by ASHRAE Standard 55-1989 (Addendum 55a-1994).

A. Building Structure

The building and site should provide effective drainage measures to control rainfall runoff and to prevent entry into the building.

The building foundation should be designed and constructed to prevent the entry of moisture and other soil gases.

Building assemblies should be designed and constructed to permit drying of interstitial spaces.

Building assemblies should be designed and constructed to prevent airflow into insulation systems from both the interior and exterior.

12 VII

Appendices

Radon resistant construction practices as referenced in the ASTM Standard "Radon Resistant Design and Construction of New Low Rise Residential Buildings" should be utilized.

Materials, adhesives and finishes with tested low emission rates should be selected.

B. Mechanical Systems

Controlled mechanical ventilation systems shall be installed.

Where combustion appliances are used, only sealed direct combustion or power vented systems should be installed in habitable spaces. Gas cooktops and gas ovens should only be installed in conjunction with exhaust fans.

Forced air systems should be designed to provide balanced airflow to all conditioned spaces and zones. Interzonal air pressure differences should be limited to 3 Pa.

Filtration systems should be provided for forced air systems which provide a minimum atmospheric dust spot efficiency of 30 percent (derived from ASHRAE Standard 52-1994).

Indoor humidity should be maintained in the range of 25 to 60 percent by controlled mechanical ventilation, mechanical cooling or dehumidification.

C. Occupant Considerations

Occupants should be provided with an operator's manual containing specific operating instructions on how to maintain a healthy indoor environment.

Control systems should include advisory display or indicative modes to alert occupants to "trouble" or "failure" conditions.

Environmental Impact Criteria

Energy efficient and resource efficient construction should be designed, constructed and operated to reduce overall life-cycle impact on the environment considering energy consumption, resource use and labor inputs in the fabrication, erection, modernization, operation and disassembly of the building, components and systems.

The design and construction of buildings should use recycled materials, or new materials with a high recycled content. Minimization of scrap on site and design for disassembly should be provided.

VII 12

Appendices

Discussion Relating to Criteria

The Criteria are for the most part self-explanatory. However, two concepts require explanation: air leakage coefficients and ductwork leakage. The following discussion relates to these two concepts.

In selecting the approach to measure/evaluate air leakage the following factors were considered:

- A single airtightness value was selected as it has become clear that it is as important to build a tight building envelope in the hot, humid south as in the cold north. Similarly, mixed, humid climates and hot, dry climates also require tight building envelopes. The importance of tight construction goes far beyond energy conservation. Health and durability are the principle concerns with respect to this issue.

- The airtightness value is based on the surface area of the building envelope not the volume. Air change per hour at 50 Pascals was rejected as a basis for measurement because it confuses the issue. We are dealing with leakage through the building envelope. Holes, holes, holes. Of course, ach @ 50 Pa is a popular, albeit misguided, criteria. The requirements have been translated for information purposes only. Based on ach @ 50 Pa, values are between 3.2 and 2.8 for 1,500 to 2,500 square foot houses with basements (not including the basements in these square footage determinations). The airtightness value is roughly double the Canadian R-2000 tightness requirement of 1.5 ach @ 50 Pa although it is roughly twice as tight as conventional construction.

Air Leakage - Determining Leakage Ratios and Leakage Coefficients

Using a blower door, measure the flow rate necessary to depressurize the building 50 Pascals. This flow rate is defined as CFM50. Alternatively, determine the Equivalent Leakage Area (EqLA) in square inches at 10 Pascals using the procedure outlined by the Canadian General Standards Board (or alternatively, determine the ELA using the ASTM procedure calculated at 4 Pascals). When determining these values, intentional openings (design openings) should be closed or blocked. These openings include fireplace dampers and fireplace glass doors, dryer vents, bathroom fans, exhaust fans, HRV's, wood stove flues, water heat flues, furnace flues and combustion air openings.

Calculate the leakage ratio or the leakage coefficient using the entire surface area of the building envelope. When determining the surface area of the building envelope, below grade surface areas such a basement perimeter walls and basement floor slabs are included.

For example, a 2,550 square foot house constructed in Grayslake, IL has a building envelope surface area of 6,732 square feet and a conditioned space volume of 33,750 cubic feet (including the basement). The measured Equivalent Leakage Area (EqLA) using a blower door is 128 square inches. This also corresponds to a blower door measured CFM50 value of 1,320 cfm and a blower door measured 2.3 airchanges per hour at 50 Pascals.

Surface Area	EqLA	CFM50	ach @ 50 Pa	Volume
6,723 ft²	128in²	1,320 cfm	2.3	33,750 ft³

To determine the Leakage Ratio, divide the surface area of the building envelope by 100 square feet and take this interim value and divide it into the EqLA.

$$6,732 \text{ ft}^2 \div 100 \text{ ft}^2 = 67.32$$

$$128 \text{ in}^2 \div 67.32 = 1.9 \text{ in}^2/100 \text{ ft}^2$$
(Leakage Ratio)

To determine the Leakage Coefficient, divide the CFM50 value by the surface area of the building envelope.

$$1,320 \text{ CFM50} \div 6,732 \text{ ft}^2 = 0.20 \text{ cfm/ft}^2$$
(Leakage Coefficient)

Many airtightness measurements are recorded as air changes per hour at a pressure differential of 50 Pascals (ach @ 50 Pa). To convert ach @ 50 Pa to CFM50 multiply the volume of the building envelope (including the basement) by the ach @ 50 Pa and divide by 60 min/hour.

For example, 2.3 ach @ 50 Pa across a building envelope of volume 33,750 ft³ is equivalent to a CFM50 value of 1,320 cfm.

$$33,750 \text{ ft}^3 \times 2.3 \text{ ach @ } 50 \text{ Pa} \div 60 \text{ min/hr} = 1,320 \text{ CFM50}$$

Ductwork Leakage

To determine the allowable limit for ductwork leakage, determine the rated air flow rate of the air handler, furnace, air conditioner, etc. at high speed from the manufacturer's literature. For example, a typical heat pump system may have a high speed flow rate of 1,200 cfm across the blower according to literature supplied with the unit. Ten percent of this value is 120 cfm. This 10 percent value becomes the total ductwork leakage limit when the total air handling system is depressurized to 25 Pascals with a duct blaster.

VII 12

Appendices

Appendix VIII
Additional Resources

Organizations

Advanced Energy Corporation
909 Capability Drive
Suite 2100
Raleigh, NC 27606-3870
(919) 857-9000
http://www.advancedenergy.org

American Council for an Energy-Efficient Economy
1001 Connecticut Avenue, NW
Suite 801
Washington, DC 20036
Research and Conferences (202) 429-8873
Publications (202) 429-0063
http://aceee.org

Energy Efficient Building Association
490 Concordia Avenue
St. Paul, MN 55103
(651) 268-7585
http://www.eeba.org

Florida Solar Energy Center
A Research Institute of the University of Central Florida
1679 Clearlake Road
Cocoa, FL 32992
(407) 638-1000
http://www.fsec.edu

VIII 12
Appendices

Appendices

. . . .

Residential Energy Services Network
    12350 Industry Way #208
    Anchorage, AK 99508
    (907) 345-1930
    http://www.natresnet.org

Rocky Mountain Institute
    1739 Snowmass Creek Road
    Snowmass, CO 81654-9199
    (970) 927-3851
    http://www.rmi.org
    http://www.natcap.org

Southface Energy Institute
    241 Pine Street
    Atlanta, GA 30308
    (404) 872-3549
    http://www.southface.org

U.S. EPA ENERGY STAR Buildings Program
    U.S. EPA Atmospheric Pollution Prevention Division
    401 M Street, SW (6202J)
    Washington, DC 20460
    (888) STAR-YES
    http://www.epa.gov/energystar.html

Publications - Books

Building Air Quality
    U.S. Environmental Protection Agency
    Indoor Air Division
    Office of Air and Radiation
    Washington, DC 20460
    (202) 564-7400

Canadian Home Builders' Association Builders Manual
    Canadian Home Builders' Association
    150 Laurier Avenue West
    Suite 500
    Ottawa, Ontario, Canada K1P 5J4
    (613) 230-3060

12 VIII
Appendices

Energy Source Directory: A Guide to Products Used in Energy Efficient Construction

> Iris Communications, Inc.
> P.O. Box 5920
> Eugene, OR 97405
> (541) 484-9353

Moisture Control Handbook

> Lstiburek, J.W. and Carmody, J.
> John Wiley & Sons, Inc., New York, NY
> ISBN 0471-318-639
> (212) 850-6306

or from

> Building Science Corporation
> (978) 589-5100
> http://www.buildingscience.com

Understanding Ventilation: How to Design, Select and Install Residential Ventilation Systems

> Bower, J.
> The Healthy House Institute
> 430 N. Sewell Road
> Bloomington, IN 47408
> (812) 332-5073

Publications - Periodicals and Catalogs

Energy Design Update

> Cutter Information Corporation
> 37 Broadway
> Arlington, MA 02174
> (800) 964-5118

Environmental Building News

> 122 Birge Street
> Suite 30
> Brattleboro, VT 05301
> (802) 257-7300
> http://www.ebuild.com

VIII 12

Appendices

Home Energy Magazine
    2124 Kittredge Street
    No. 95
    Berkeley, CA 94704
    (510) 524-5405

Journal of Light Construction
    932 W. Main Street
    Richmond, VT 05477
    (800) 375-5981

Solplan Review
    Box 86627
    North Vancouver, British Columbia, Canada V7L 4L2
    (604) 689-1841

Index

A

Air Drywall Approach (ADA) 253, 257-259
air flow retarder. **See** polyethylene
air handler 8, 14, 23, 128, 136, 137, 145-148, 150-153, 155, 157, 159, 302, 313, 314, 317
air-to-air heat exchanger 138
air-to-air heat pump 141
aluminum siding. **See** siding:aluminum
appliance(s) 3, 6, 8, 14, 23, 24, 150, 152, 303-305
   combustion 4, 24, 129, 130, 143, 307, 310, 314, 315
attic 8, 20, 23, 38-42, 146-148, 161, 163, 164, 168, 176, 178, 304

B

baffle 176, 177
band joist 56, 57, 161
barbecues
   indoor 24, 148, 304
basement 34, 36, 37, 41, 42, 44, 58-63, 65, 66, 68, 129, 146, 147, 191, 310, 316, 317. **See also** concrete
blower door 300, 301, 316, 317.**See also** fan: calibrated
brick 45, 49, 51, 62, 63, 238, 240, 264, 272, 287
building envelope 3, 6, 20, 21, 23-25, 128-130, 148, 152, 165, 229, 251, 261, 265, 266, 297, 299-302, 304, 309, 311-313, 316, 317
building paper 2, 50, 63, 161, 228, 236, 237-240, 251, 252, 256, 260, 263, 264, 269, 271, 275, 277, 279, 282-284, 287

C

carbon dioxide 181, 182
carbon monoxide 305
carpet(s) 3, 35, 36, 146
ceiling 168, 178, 183, 184, 186, 188
   cathedral 171, 177
   dropped 152, 155, 165
cement 51, 60, 63, 181
chimney 3, 17, 20, 182. **See also** fireplace
commissioning 14, 24, 308, 311
concrete 22, 31-33, 35, 43-45, 48-59, 61-66, 145, 227, 261, 264
control joints 31, 36, 180. **See also** concrete
crawl space 3, 13, 20, 23, 34, 35, 39-41, 50-56, 68, 145-147, 150, 163, 168, 173, 263, 303
cross bracing 29

Index

Index